Microbiology: Laboratory Manual for Allied Health and General Microbiology

Jay F. Sperry

University of Rhode Island

KENDALL/HUNT PUBLISHING COMPANY
4050 Westmark Drive Dubuque, Iowa 52002

ISBN 978-0-8403-9622-8

Printed in the United States of America
15 16 17 18 19 20 21

Contents

PREFACE

This microbiology laboratory manual has been used for many years for both a general introductory microbiology laboratory manual and for introductory medical microbiology laboratory manual for the allied health disciplines at the University of Rhode Island. It gives a comprehensive coverage of the basics of introductory microbiology, plus some coverage of its applied aspects. Sufficient detail has been given so that the exercises may be performed independently.

Numerous instructions have been provided so that the exercises may be perfomed with a high degree of success. The appropriate steps have been included, as well as directions concerning when observations should be made and what the results should be. We have also included a number of questions at the end of most exercises. This will allow you the opportunity to challenge your understanding of the principles and applications of the information obtained from the experiment.

You will be learning the fundamentals: safety in the microbiology laboratory, aseptic tecchnique - how to handle microorganisms safely and transfer them between vessels without contaminating either yourself, others in the laboratory or the cultures of microorganisms you are working with. You will also be learning where microbes are found in and around us as well as how to culture them and identify them.

INTRODUCTION

The laboratory course in introductory Microbiology has several objectives. First, it is intended to give the student some acquaintance with bacteria and other microorganisms. Secondly, the experiments are designed to illustrate various phases of microbial life and activity, and in so doing to assist in understanding biological problems of a fundamental nature. Thirdly performance of the work provides practice in the techniques of handling microorganisms. Whether or not the student plans to study microbiology further, mastery of certain technical procedures will be of value, either as a foundation for more advanced work or be emphasizing the fact that microorganisms constantly surround us and must be considered in our daily life.

Emphasis will be placed upon development of technique and interpretation and understanding of results. The textbook should be consulted for explanations, and if these appear inadequate, ask the instructor. Examinations will frequently demand reasoning and interpretation of facts, as well as the facts themselves.

The general plan of the laboratory plan will be: (1) a microscopic study of microbial morphology; (2) study of the physiology of microorganisms; (3) study of environmental effects on microorganisms; (4) isolation and identification of unknown bacteria and pathogens; (5) study of some practical situations in which bacteria are significant in health and medicine.

Laboratory notes may be kept in a separate notebook, preferably a spiral bound book. A separate page or more should be allotted to each experiment and notes should be written up in the laboratory, as observations are made. Do not jot results on a scrap of paper and then write them up later. Be brief, but give sufficient detail so that the notes will mean something to you and to the instructor who may review the book. Use this general outline:

(a) Number and title of experiment
(b) Object of the experiment
(c) Procedure (a brief summary or pasted in Xeroxed directions)
(d) Results (include those of neighbors if do directed)
(e) Conclusions

It may occasionally be necessary to request members of the class to come to the laboratory for a few minutes at their convenience on a day when no regular period in scheduled, in order that they may have bacterial cultures of proper age for later use, or to make observations on experiments at, for example, 24 hours after starting them.

The laboratory work, textbook assignments, lectures and recitation discussions will cover aspects of the same topics as far as schedules will permit. Recitation quizzes and hour exams will frequently contain questions relating to the laboratory experiments.

Laboratory Rules and Regulations

1. Hang coats on hangers and put books on the shelves provided.

2. Do not bring food into the laboratory.

3. Smoking and gum chewing are not permitted.

4. Learn immediately to keep paper, pencils, fingers, and other objects out of the mouth.

5. Keep your work area neat and orderly at all times.

6. Wash the bench top with disinfectant solution before starting to work and after completing the days exercises.

7. Discard match sticks, paper, etc., in the pans provided or in the wastebaskets.

8. Discard your cultures in the baskets and trays provided. Leave the laboratory cultures at the desk where you find them.

9. The needle or loop used in transferring cultures must be sterilized in the flame <u>before and after use</u>. Adjust the burner so that a blue flame is obtained. Hold the needle or loop <u>vertically</u>, not horizontally, about 1/4 inch above the point of the blue flame. If the needle or loop is covered with viscous material or an excessive amount of any material, <u>dry it at the side of the flame before sterilizing it</u> to avoid spattering and scattering living material.

10. Learn immediately that many of the organisms with which you will be working are, or can be, dangerous.

11. If a culture is spilled, notify the instructor.

12. Do not hoard materials. When an experiment is finished and the results have been recorded, discard cultures and the materials used.

13. Handle the microscope with care. Use two hands, one to carry the microscope, the other to assist in lowering it to the desk top or in replacing it in the microscope cabinet.

Avoid Jolting and Jarring The Instrument

14. When you have finished using the microscope, clean excess oil from the oil immersion lens with lens paper (never use cheese cloth or paper towel). Turn the revolving nosepiece so that the low power lens is down before returning the microscope to the cabinet.

15. Turn off the Bunsen burner before examining the preparations under the microscope. There is danger of catching your hair on fire if you are working close to a flame. Long hair should be tied up.

16. Results of experiments are to be recorded directly in your laboratory notebook. Do not record results on a piece of paper with the idea that they will be transcribed later.

17. Avoid unnecessary conversation, noise or commotion.

18. Read directions. Listen to instructions. Follow them. Laboratory comments will be given promptly at the beginning of the laboratory period. These directions do not preclude your reading the directions in the lab manual either before the laboratory period or before you begin work. Be at your desk for these instructions when the class starts. Talking, arranging of materials, labeling of cultures, etc. will not be tolerated during this instruction period.

19. Included with oral direction will be discussion of some aspects of the experiments. It is suggested that notes be taken because some of the information may not be found in your textbook (or perhaps any other). You will be held responsible for it on quizzes and examinations.

20. Your laboratory grade depends upon many factors.

    a. Technique. Bacteriological technique is relatively simple. It is realized that individuals differ in manual dexterity and that many will not master the technique in the short time allowed in a beginning course. The student without dexterity; but, who makes who makes a conscientious effort to develop technique will not be penalized. In actuality he/she may receive a better mark than the student who has the ability but refuses or neglects to use it.

    b. Comprehension of the experimental design and of the principles underlying each experiment.

c. General class performance, including conduct, listening to instructions, following directions, attention to details such as neatness, orderliness, care of microscope, etc.

d. Inspection of laboratory notebooks may or may not be done.

e. The isolation and identification of an unknown microorganism.

EXPERIMENT 1

SMEAR PREPARATION

Most bacteria are difficult to see in the light microscope if they are unstained. They are stained to aid in their classification and to identify structures such as endospores, flagella, capsules and granules. Prior to the staining process, the bacteria are usually attached to a glass microscope slide.

A. **Slide cleaning.** Beginning students usually find it difficult to make bacteria stick to the slide. The problem usually is the result of greasy slides rather than the drying or heat fixing process. Do not use hand soap to clean slides. Bon ami is supplied for this purpose. After washing the slide in hot water with Bon ami, the slide is often made greasy again by fingers. Holding the slide by the edges, quickly flame both sides of the slide in the Bunsen burner.

B. **Loop size**. Broth is transferred with an inoculating loop to a single spot on the slide. The loop, if properly prepared, should be able to hold a drop of liquid. The drop will not hold if the loop is too large or too small. The following procedure is recommended for preparing a loop of the proper size:

1. Straighten out the loop to make a straight wire.

2. Bend it around the handle of another inoculating loop so that the wire forms a closed ring. Then slip it off the handle.

3. With two fingers, grasp the loop and bring it to an upright position. Make sure the loop ring is closed tightly.

C. **Loop flaming.** Always flame the loop before and after using it. The length of the wire should glow red hot. The handle is not flamed and should not be inserted into a sterile test tube where it could touch the inner side of the tube. The tube should be held in a slanted position so that dust does not fall from the air or from the handle of the inoculating loop.

D. **Test tube flaming**. To help prevent the dust and unwanted bacteria from entering the tube, the lip of the tube should be flamed immediately after the plug is removed. The flaming process serves: (1) to kill bacteria that are on the lip of the tube and (2) to heat the tube so that expanding hot air will convect dust away from the entrance. The tube is flamed again after the sample has been removed, and the cotton plug or cover is replaced.

E. **Procedure for making smears of bacteria from broth cultures.**

1. transfer a loopful of broth to a clean slide. Flame the loop before and after each time it is used.

2. Spread the broth over an area about the size of a nickel.

3. Allow the smear to dry and then heat fix by very quickly passing the slide once or twice through the Bunsen flame or by heating it over a microscope lamp.

F. **Procedure for making smears from colonies on agar.**

1. Place a loopful of tap water on a clean microscope slide. Flame the loop before and after each time it is used.

2. Flame the inoculating loop or needle, let it cool 5 seconds.

3. Touch the center of the colony with the tip of the needle or loop. Remove a very small amount of growth, an amount no more than just barely visible. Mix and spread the bacteria in the drop to a smear about the size of a nickel (beginning students usually put on too much). When the smear has dried, it can be heat-fixed.

G. **Fixing the smear.** Most bacteria will attach to a clean microscope slide upon drying and will remain attached throughout the staining and washing process. Bacteria with a large amount of lipid material in the outer membrane may not adhere well. These can be made to adhere to the slide by slightly heating the dried smear either by passing the slide through the Bunsen burner once or twice to warm the slide, or drying the slide over the microscope lamp. This process is referred to as "heat-fixing". This process should be done with caution. Most students heat the slide too much. Heat cooks the bacteria to a smaller size and will sometimes burst the bacteria.

Caution. Be careful in handling the inoculating needle or loop. Do not allow it to vibrate when contaminated with bacteria. The process will aerosol bacteria throughout your work area.

EXPERIMENT 2

SIMPLE STAIN PROCEDURE

Bacteria are readily stained with basic dyes such as methylene blue, crystal violet or safranin. The basic stains bind to nucleic acids (DNA and RNA) inside the bacteria. When only one stain is used the staining procedure is simplified and referred to as a "simple stain" procedure.

Procedure:

1. Mark a clean slide in thirds with a wax pencil. Label the sections so that you can identify the source of bacteria for each section.

2. Prepare smears as directed in Exp. 1. Put a sample from tube A in section 1; from tube B in section 2 and samples from A and B in section 3.

3. Place the slide on a staining rack and cover the smears with crystal violet solution. Allow it to stand for 30 to 60 seconds.

4. Decant off the dye and gently wash with tap water. Hold the slide at an angle and allow the water to gently rinse the smears. Do not allow the water to hit the smears directly, it may wash off the bacteria.

5. Shake off excess water and gently blot the smear dry by pressing it between bibulous paper or a paper towel.

6. Mount the slide on the microscope. Observe the smear under low power, high power, then with the oil immersion objective. The latter will require the use of immersion oil. Add a drop of oil directly onto smear. The oil immersion objective is lowered into the oil until it gently touches the slide. Bring the condenser up about as far as possible. Open up the iris diaphragm enough for sufficient light.

7. Slowly focus upward with the course adjustment. When the purple spots appear, finish focusing with the fine adjustment. When the bacteria are in focus, consult page for proper microscope alignment.

8. Sketch a few of the bacteria, showing there shape and arrangements.

BRIGHT FIELD MICROSCOPY

In bright field microscopy, light passes unaltered through the lens system to produce a brightly lighted field.

Magnification. The limit of magnification for bright field microscopy is 1,000 diameters. Your microscope contains an ocular (eye piece) which will provide a ten-fold magnification. The objective lenses of your scope are 10X (low power), 45X (high dry) and 97X (oil immersion). These when combined with the magnification of the ocular will provide 100-, 450- and 970-fold magnification respectively. Each lens has a different working distance from the specimen. The distances for the 10X, 45X and 97X objectives are 8.3 mm, 0.72 mm and 0.14 mm respectively. The objectives are usually set in the nosepiece so that if one objective is in focus, the other objective will automatically be in focus when rotated in position. This is referred to as parfocalling. It is usually easier to find a specimen using the high dry objective and to swing the oil immersion objective in place. The millimeter markings on the objective lens refers to focal length.

Resolving power. The ability of the lens system to distinguish between two objects or lines that are close together is called resolving power. It is related to the wavelength of light by the following formula:

$$\text{Resolving power} = \frac{0.61 \times \text{wavelength}}{\text{numerical aperture}}$$

The shorter the wavelength of light the greater the resolving power. These shorter wavelengths are found in the blue-violet region. Longer wavelengths interfere. To reduce the interference, a blue filter is passed between the light source and microscope slide.

Iris diaphragm. The resolution and sharpness of image also depends on the settings of the iris diaphragm, condenser and use of immersion oil. A diaphragm that is open too far reduces the depth of the field that will be in focus; whereas, closing the diaphragm too much decreases the numerical aperture (the mathematical expression for the cone of light delivered to the specimen by the condenser and taken up by the objective).

Condenser. The condenser serves to bring the light rays into focus of the specimen. When using the oil immersion objective, the condenser is adjusted to give maximum brightness. Raising or lowering the condenser to provide less than maximum brightness reduces the numerical aperture and consequently the resolution.

Immersion oil. Immersion oil has the same refractive index as glass. It serves to prevent the loss of light between the specimen and objective lens. Without immersion oil a large amount of light would be lost because light that is refracted by the air would miss the objective lens. This loss brings about a decrease in the numerical aperture and in turn resolution. Since the iris diaphragm and the condenser is adjusted to provide the desired resolution, it should not be changed to reduce the light intensity when using the oil immersion objective.

CARE AND USE OF THE MICROSCOPE

General Care of the Microscope

The primary rule is to keep the microscope free of dust. Repairmen recommend that the scopes be covered with plastic bags even when stored in cabinets.

Treat the scope gently and avoid sudden jars. When transporting the microscope, hold it upright and use both hands, one under the base and the other grasping the arm of the scope.

Care of Objectives and Eyepieces

Lens paper is provided for cleaning lenses. Fingers should not come into contact with any of the lens surfaces. Do not take apart the objective lens. Do not force the microscope. Notify the instructor if the microscope is not functioning properly. Clean immersion oil off the objective lenses immediately after use. Failure to do this will loosen the lens, making it inoperative (cost of repair - approximately $150). Do not clean objective lens with xylol. This is a delicate procedure and should be done only by the instructor.

Focusing and Adjustment of Diaphragm and Condenser Lens for Use with the Oil Immersion Objective.

Improper settings of the diaphragm and condenser will bring about a loss of definition of the object. For general laboratory usage, Bausch & Lomb Co. recommends the following procedure:

1. Focus the microscope on the object (use oil immersion objective).

2. Remove eyepiece (ocular), open diaphragm (wide open).

3. Look down body tube and adjust the condenser to provide maximum amount of light in the tube. This setting is near the top or maximum condenser height.

4. Close the diaphragm until it cuts out the light by about one-third. At this point replace eyepiece, slightly adjust the diaphragm until the best vision is obtained. Under no circumstances should the diaphragm be open wider than is sufficient to fill the objective with light.

5. Always focus upward. Do not look through the microscope while lowering the tube with the course adjustment. Do not exchange objective or eyepieces with scopes of other brands or types.

6. Store scope with low power objective in place, adjust mechanical stage so that it does not extend beyond the edge of microscope stage, clean off all oil, giving special attention to the objective and condenser lenses.

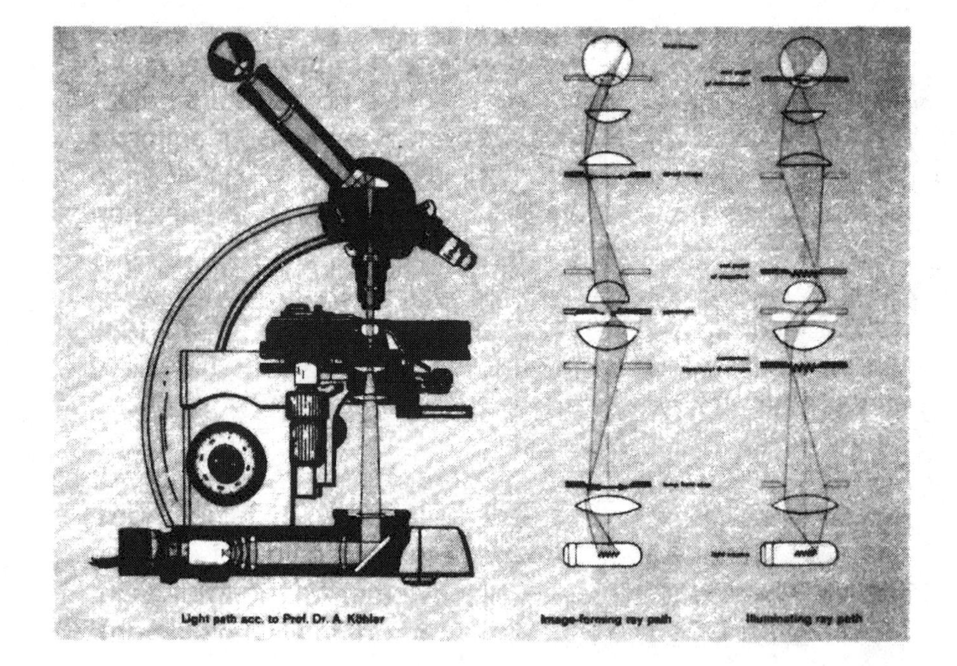

Courtesy of H. E. Keller, Carl Zeis, Inc., Thornwood, NY.

EXPERIMENT 3

SEPARATION AND ISOLATION OF BACTERIA FROM A MIXED CULTURE

A practical method for separating bacteria involves spreading the bacteria on an agar plate by the streak method. This method of streaking an agar plate is a form of diluting the original sample. The colonies that develop after incubation are then examined for purity.

Procedure: 1. A mixture of two kinds of bacteria will be provided.

2. Obtain a Petri plate of nutrient agar and streak with a sample of the unknown mixture in the manner described by the instructor or as described below. Use streak plate method 1 or 2.

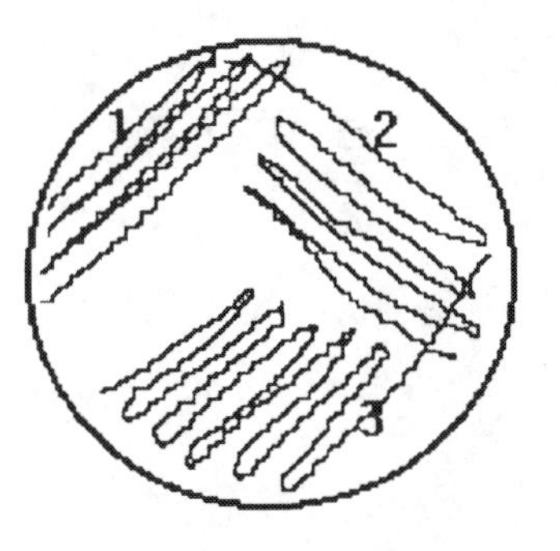

a. Transfer culture by a sterile loop to agar plate. streak 1/3 of plate flame and cool the loop by touching agar

b. Rotate the plate 90° and spread out bacteria by streaking 1/3 of the plate. Flame and cool loop.

c. Rotate the plate 90° and streak the remainder of the plate.

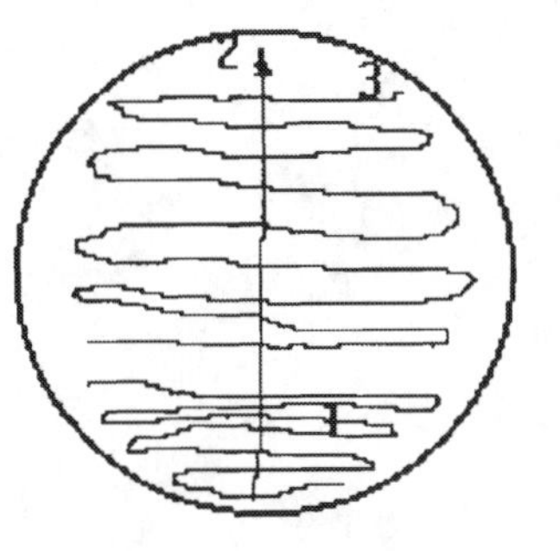

a. Streak 1/4 of the plate, flame loop, touch to agar

b. Make a single streak , through streaked part and across the plate. Flame loop and cool as before.

c. Make a continuous series of streaks across line 2, starting at the top.

It is recommended that the loop be held at the balance point. The loop should not cut into the agar. <u>Always flame loop before and after using it.</u>

3. Incubate the plate in an inverted position at 30° C until the next period.

4. Examine plate and note the different kinds of colonies.

5. Gram stain and observe with a microscope bacteria from each different type of colony. Use 100-, 450- and 970-fold magnification.

EXPERIMENT 4

THE GRAM STAIN

The Gram staining procedure is one of the most widely used in bacteriology. It is a differential stain based upon the chemical and the physical structure of the bacterial cells. Following primary staining with crystal violet and mordanting (complexing) with iodine, the cells are treated with a decolorizing agent and are counterstained. Gram positive bacteria retain the primary stain through the decolorizing step, while Gram negative bacteria lose it. It seems likely that a complex of crystal violet and iodine is trapped within Gram positive cells as a consequence of decreased cell wall penetrability, because they contain many layers of cross-linked peptidoglycan. Gram negative bacteria contain high concentrations of lipids in their surface structures, and their solubility in alcohol may contribute to the ease with which the crystal violet-iodine complex is extracted.

Procedure:

1. Prepare smears of *Escherichia coli*, *Bacillus subtilis*, or *Micrococcus luteus* and any unknowns.
2. Cover smears with crystal violet for 1 minute; then rinse.
3. Cover with Lugol's iodine for 1 minute; then rinse.
4. Decolorize with 95% alcohol for 10 seconds; then rinse.
5. Repeat step 4.
6. Counterstain with safranin for 30 seconds; then rinse, blot dry and examine using the oil immersion objective.
7. Tabulate the colors and Gram reactions (positive and negative) of the bacteria tested. What is the color of a Gram positive organism? What is the color of a Gram negative organism? What is the color of a Gram variable organism?

Questions:

1. What is the purpose of iodine in the Gram stain?

2. Why do Gram positives from old cultures sometimes stain Gram negative?

3. What factors may cause variation in the Gram reaction?

4. Of what practical use is the Gram reaction?

EXPERIMENT 5

BACTERIAL MOTILITY AND CLUSTER ARRANGEMENT
(Hanging Drop Method)

Some, but not all, bacteria are able to swim. Their appendages of locomotion are flagella. These cannot be seen unless specially stained. possession of flagella can usually be deduced; however, from the motility of an organism. Young cultures should be used, since loss of locomotion may occur after a culture is over 24 hours old. Cluster arrangement of cocci is best observed by hanging drop method.

Procedure:

1. Spot a small amount of Vaseline at each corner of a coverslip.

2. In the center of the coverslip, place 2-3 loopfuls of a young (24 hours or less) broth culture of *Bacillus subtilis*.

3. Place the hollow part of a depression slide over the coverslip so that the latter sticks to the slide.

4. Invert the slide so that the coverslip is on the top and the drop of culture hangs in the depression of the slide.

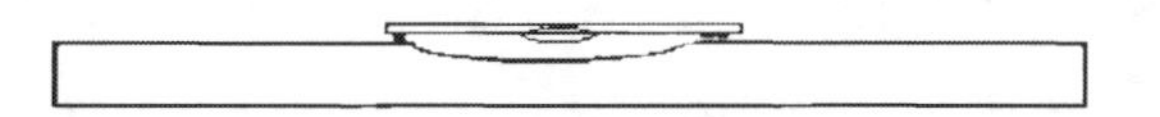

5. Observe under high dry and then with oil immersion objective of the microscope. Reduce the light with the iris diaphragm until the field is gray like a cloudy sky. Focus carefully upward. Usually it is easier to find the bacteria by focusing on the edge of the drop of culture.

6. Pick out one organism and follow it to see its path of motion. Make a sketch illustrating this by dotted lines and arrows.

7. Repeat with any unknown organisms you have (you will need a broth culture for this). Observe cluster arrangements.

Questions:

1. Distiguish between true motility and Brownian movement.

2. Do all motile bacteria have the same number of flagella?

3. What does the word peritrichous mean?

4. Is there any correlation between cell shape (rod, coccus) and motility?

5. May a motile organism temporarily lose its power of locomotion?

EXPERIMENT 6

BACTERIAL ENDOSPORES

Bacterial endospores are resistant bodies produced by certain bacteria as a stage in their life cycle. Their formation is found mainly in rod-shaped bacteria. They are more abundant in old than in young cultures. They may be demonstrated microscopically by staining procedures that make use of their resistance to the penetration of chemicals, including dyes. They may also be observed using phase conrtrast microscopy without using dyes.

Procedure: A. Simple Stain

1. Prepare a smear of *Bacillus subtilis, B. cereus,* or *B. megaterium* (2-5 day, 30°C, nutrient agar slant culture), air-dry and fix with heat.

2. Stain with crystal violet for 15 seconds, wash, dry, and examine under oil.

3. A cell containing an endospore is called a sporangium. After the spore has matured, the rest of the sporangium may autolyse, leaving the spore free. What is the color of the spore? What is the color of the vegetative cell?

4. Sketch a few vegetative cells, sporangia and free spores (if present). Are the sporangia swollen? In what part of the cell is the spore located? Is the spore oval or spherical?

B. Schaeffer and Fulton Method

1. Prepare another smear of the above mentioned bacteria.

2. Cover with malachite green; steam for 2-3 minutes (don't let the stain dry out). If you don't have a steam cabinet, the slide can be put onto a staining rack and a burner can be tilted and the stain can be heated intermittently, just enough to cause vapors to rise from the stain.

3. Rinse off excess stain by washing in tap water 30 seconds.

4. Counterstain with safranin for 30 seconds. Wash, dry and examine under oil.

5. What color are the spores? The vegetative cells?

6. Sketch and answer questions for item 4 above.

Questions:

1. Explain why spores and vegetative portions of sporangia stained as they did in the two methods employed?

2. Do bacteria normally have more than one edospore?

3. What are the common genera of bacteria which produce endospores?

4. What are some of the conditions necessary for bacteria to sporulate?

5. Where in a bacterium may endopsores be found?

6. How does spore formation in bacteria differ form sporulation in yeasts and molds?

7. What is a differential stain?

8. What is a counterstain? What is its purpose?

EXPERIMENT 7

PREPARATION AND STERILIZATION OF CULTURE MEDIA

Media used for cultivation of bacteria will vary in composition according to the nutritional requirements of the bacteria being studied. Media used for the first part of this course will be those traditionally used for the most common types of bacteria. These media are grouped into two categories: (1) nutrients suspended in broth and (2) broths to which solidifying agents have been added. The latter are referred to as "agars" (e.g. agar tubes, agar plates). The solidifying substance (agar) is a polysaccharide of galactose obtained from a red agal seaweed. Bacteria studied in diagnostic laboratories are unable to digest agar, making it an ideal solidifying agent. Agar dissolves at 100°C and solidifies at 42°C.

The most commonly used broth is nutrient broth. It is composed of peptone (an enzyme or hydrochloric acid digest of protein, in this case gelatin) which contains various length peptides that can be transported across bacterial membranes. Some insoluble proteins, like casein, are converted to soluble peptones for use in bacteriological media. Beef extract is added to provide vitamins and minerals.

Procedure: A. Preparation and sterilization of media

1. Work in groups of 4. Mix ingredients for 200 ml of broth.

peptone	1.0 g
beef extract	0.6 g
water	100 ml

2. Dispense 100 ml of broth in 5 ml aliquots to test tubes, and cap tubes with morton closures, or plug with cotton.

3. Add 1.5 g of agar to the remaining broth, and bring to a boil with stirring, to dissolve the agar.

4. Warm a 10 ml pipet in your Bunsen flame and dispense the agar in 5-6 ml aliquots to test tubes and cap tubes.

5. Autoclave the broth and agar tubes at 121°C for 15 min.

6. After autoclaving, slant the tubes of nutrient agar at a very shallow angle.

EXPERIMENT 8

PHYSIOLOGICAL CHARACTERISTICS OF BACTERIA

The following series of experiments is designed to provide insight into the approach we use to identify unknown microorganisms and to provide some information concerning their nutrition and physiology. The techniques are standard and will not only help to understand concepts, but will be of value if you become employed in a microbiology laboratory. While doing these experiments, it is important that you observe every detail and ask yourself many questions. You should be comparing colonies and biochemical reactions of known microorganisms to your unknown microorganism. It is also important to talk with your fellow students about their unknown organisms.

Step I. Isolation of the microorganism in pure culture

Use the streak plate method (Experiment 3) for obtaining your unknown in pure culture. Do this carefully, but quickly to avoid contamination. Each time that you contaminate your unknown, you will have to reisolate it and start it over. At the next laboratory session, observe the colonies on the streak plate and describe them in your lab notebook. Then inoculate two nutrient agar slants with your unknown which you have taken from an isolated colony on a streak plate. Incubate the slants at 30°C until the next lab session. After the bacteria have grown on the slant, Gram stain (Experiment 4) a sample and the label the slant (*your name, date, unknown)* and present it to your laboratory assistant for storage between laboratory sessions.

Colonies will often give clues to help categorize your unknown. Observe the **colony** carefully and try to determine the following:

a. Is the colony opaque or translucent (light transmission)?

b. Is the surface smooth, wrinkled, concentric or contoured?

c. Is the elevation very flat (spread thinly over the plate), raised or convex?

d. Is the shape circular, punctiform (pin point), irregular or rhizoid (having root-like outgrowths)?

e. Is the margin (edge) entire (straight), undulate, irregular, filamentous or curled?

f. Does the colony have an odor (earthy, fruity, ammonia)?

g. Is the colony pigmented (colored)? Does the colony change the color of the medium?

h. Gram stain a sample of the colony that was used for inoculating the two slants; record your results.

Step II. Gram stain your unknown bacterium

After you have prepared a streak plate, carefully prepare a smear of your unknown and Gram stain it. Until you really known the Gram reaction, shape (coccus, rod or spiral) and the arrangement (single, chain, cluster, diplo, tetrad) of your unknown; you will not be able to start the identification process. You will also be unable to tell if you unknown has become contaminated.

Step III. Motility

Most cocci are non-motile. You still have to be sure that you don't have an exception. Motile Gram (-) rods are classified by their flagellar arrangement. Beginning students find the flagella stain exasperating and they do not have time to develop the technique properly. Therefore, we make a rough judgment for polar flagellation by observing the swimming characteristics of the unknown in a hanging drop preparation (Experiment 5). If the movement is markedly "fish tail", we conclude that the unknown is polarly flagellated.

Motility is best observed when bacteria have sufficient energy. At maximum stationary phase of growth, the energy source may have become depleted. Select young cultures. If your unknown is a rod and apparently nonmotile at 48 hours, re inoculate and observe it when the culture is actively growing (18-24 hr).

Step IV. Respiration or fermentation? (mostly for Gram-positive cocci)

After the unknown is characterized by shape, staining reaction and special anatomical features such as spore or flagellar arrangement, the unknown is classified further by its metabolic reactions. The next step is to determine whether it has a metabolic system to allow it to grow only as a strict aerobe (requires molecular oxygen) or as a facultative anaerobe (grows with or without molecular oxygen); also if it catabolizes glucose anaerobically.

Oxidation-fermentation (OF) tubes.

The purpose of this test is to determine the catabolic pattern your unknown uses on glucose, whether your unknown can catabolize glucose. You will inoculate two tubes of OF glucose media by stabbing each tube with a loopful of your unknown, then overlay the medium in one tube with sterile mineral oil (this overlaying step must be done aseptically). Incubate the tube at 30°C and observe after 48 hours.

S. epidermidis cannot hydrolyze starch

Motility - non-motile
OF tube - anaerobic
✓ did not hydrolyze starch
hydrolyzed gelatin
hydrolyzed casein
{ glucose fermented
lactose not fermented
sucrose not fermented }
production of nitrite. (anaerobic)
broke down tryptophan → indole
H_2S not present. did not break down cysteine

+ urease production
− oxidase
can use glucose, sucrose, lactose to form acid products
cannot hydrolyze gelatin

Brief conclusion on the culmination of your bacteria & expected name of your bacteria. Consistent or not-so-consistent results & what the problem might have been. contamination? False positive? Multiple experiments that tested the degree of anaerobic or aerobic capacity. Thorough, yet concise. Should be two-three paragraphs.

Endospores - usually rods
H_2S - probably not

Experiments we did:

Gram stain - Exp. 4
Motility - Exp. 5
Endospores - Exp. 6
Physiological characteristics - Exp 8 (OF tube)
Hydrolysis of starch - Exp 9
Hydrolysis of Gelatin & Casein - Exp. 10
Fermentation of carbohydrates - Exp 11
Reduction of nitrates - Exp. 12
Production of indole Exp 13
Production of H_2S - Exp 14
MR-VP citrate test - Exp 16

Characteristics

- heterofermentative
- catalase negative
- facultatively anaerobic
- produced acid from glucose, fructose, mannose, ribose, cellobiose, trehalose, and from salicin, but not from sucrose & lactose
- pH less than 4 & in 10% CH_3CH_2OH

Strict aerobes. Bacteria that are strictly oxidative will grow only at the top of the medium in the tube without the mineral oil overlay. Acids may or may not be produced. If no acid is produced, glucose is converted to CO_2 and H_2O, the pH indicator will remain green. If acids are produced, the indicator will turn yellow. If the bacterium is incapable of catabolizing glucose it will catabolize peptone in the medium, release ammonia and cause the indicator to turn blue (these results will usually indicate you have an oxidase positive, nonfermentative Gram negative rod).

Facultative anaerobes. The facultative anaerobes will turn both OF glucose tubes yellow. If your unknown is a coccus, produces catalase and turns both OF glucose tubes yellow it should be a member of the Genus *Staphylococcus*.

Step V. Biochemical Reactions

Experiments 9 through 19 provide the chance to observe various metabolic activities of bacteria and the differences among bacteria. These variations are the results of different genotypes and can be used as a tool for classifying bacteria. Caution must be exercised in interpreting the results of these tests. For example, your observations using Durham tubes (fermentation tests) should correlate with those made from the glucose OF test (Step IV). A strict aerobe can produce acids which may diffuse from the surface throughout the tube and therefore appear to produce a fermentative reaction. Often the acids are distributed throughout by agitation while handling the tubes. Therefore, you cannot always conclude that fermentation has taken place just because the tube has turned yellow. You must correlate your results with your knowledge of its respiratory abilities.

In each experiment, be careful to observe whether or not growth has taken place. For example, in the nitrate test, do not add chemicals to tubes that show no growth. Where no growth is observed, re inoculate with a fresh inoculum. A strict aerobe may require tubes containing only a small amount of broth (shallow tubes).

Dichotomous keys are included to help you to identify your unknown bacteria. First, you must know the Gram reaction and shape to get to the proper key. These keys include some additional test reactions: casein hydrolysis, catalase, oxidase and urease tests.

EXPERIMENT 9

PRODUCTION OF EXTRACELLULAR ENZYMES HYDROLYSIS OF STARCH

Certain complex food molecules are unable to diffuse into a bacterial cell before being hydrolyzed into smaller molecules. Some bacteria produce extracellular enzymes that catalyze the hydrolytic reactions, often at some distance from the cells. Starch is an example of a complex carbohydrate that may be digested in this way. A test for the ability to hydrolyze starch is useful in helping identify some species of organisms.

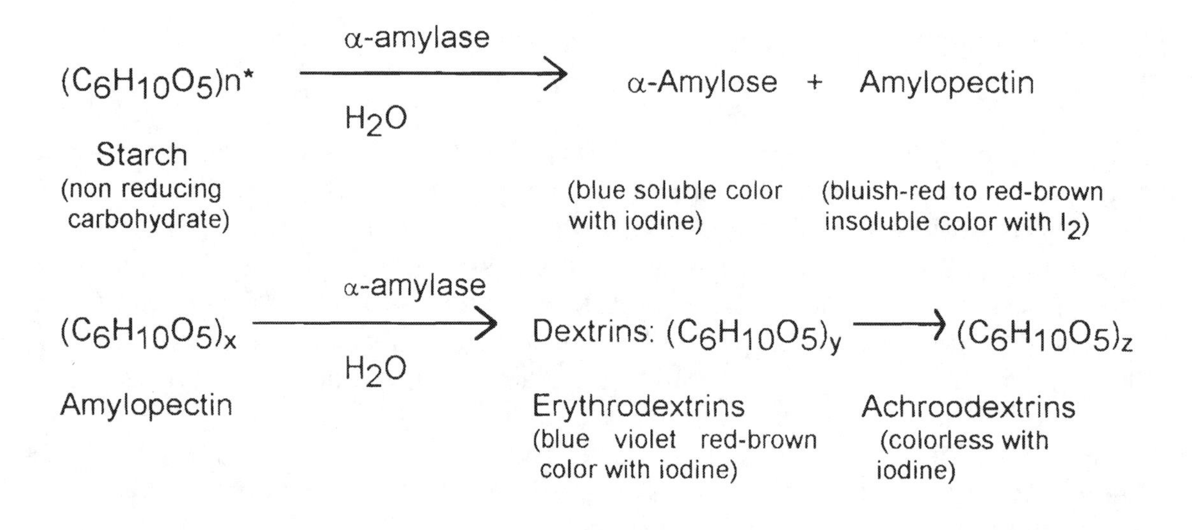

Procedures:

1. Mark a plate of starch agar into as many sections as you have organisms to test.

2. Inoculate each section in a single test spot with a loopful of one test culture. Include *E. coli, B. subtilis* and any unknowns you have.

3. Incubate the plate at 30°C until next period.

4. Flood the plate with Lugol's iodine and pour off the excess.

5. sketch the appearance of the plate, indicating areas in which starch is still present by an appropriate shading.

Questions:

1. What evidence is there in this experiment that the starch-hydrolyzing enzyme is extracellular?

2. What is the name of the starch-hydrolyzing enzyme?

3. What type of activity is usually characteristic of extracellular enzymes?

4. Does hydrolysis of starch benefit the organism? Explain.

5. Is it theoretically possible to prepare a bacteria-free preparation of an enzyme such as amylase by filtering a culture of an amylase-producing organism? Could all of the enzymes of bacteria be secured by such process?

EXPERIMENT 10

PRODUCTION OF EXTRACELLULAR ENZYMES
HYDROLYSIS OF GELATIN AND CASEIN

Gelatin is used in this experiment to demonstrate extracellular hydrolysis of protein. When gelatin is hydrolyzed, it loses its ability to form a gel or to precipitate when treated with a protein coagulant.

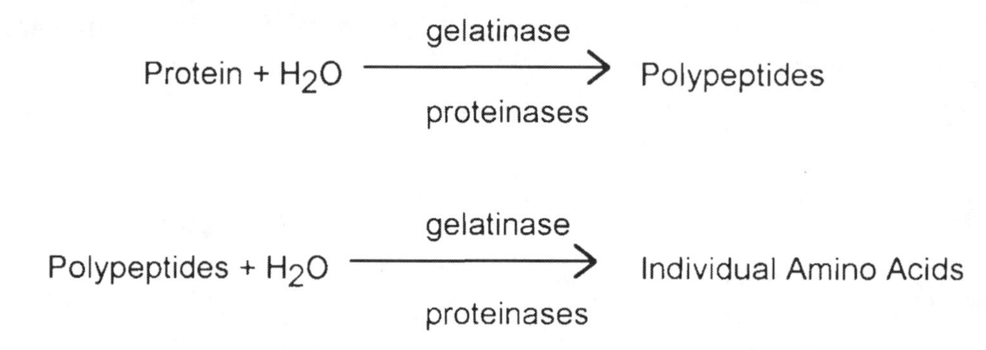

$$\text{Protein} + H_2O \xrightarrow[\text{proteinases}]{\text{gelatinase}} \text{Polypeptides}$$

$$\text{Polypeptides} + H_2O \xrightarrow[\text{proteinases}]{\text{gelatinase}} \text{Individual Amino Acids}$$

Procedure: A. Frazier's Plate Gelatin Agar.

1. Mark a plate of Frazier's gelatin agar into as many sections as you have organisms to test.
2. Place an inoculum in the center of each section and incubate the plate at 30°C. Include *E. coli, B. subtilis* and any unknowns you may have.
3. Test for undigested gelatin by flooding the plate with acid (20% HCl), letting it remain 10 - 20 minutes. Hydrolysis of gelatin is shown by a clear zone surrounding a colony.

B. Casein Hydrolysis

Casein agar consists of nutrient agar and 1% skim milk. The casein gives the agar a milky opalescence which will clear upon hydrolysis.

1. Follow steps 1 and 2 as described for the Frazier's gelatin plate. Look for clear zones around the bacterial growth.

Questions: How does this experiment demonstrate that gelatinase and caseinase are extracellular enzymes?

What are the products of protein hydrolysis?

EXPERIMENT 11

FERMENTATION OF CARBOHYDRATES

Bacteria differ greatly in their ability to ferment various carbohydrates. Because of this, fermentation tests are useful in helping to classify bacteria. the products of fermentation include: acids, alcohols and gases. Not all bacteria produce all these kinds of waste-products. Acids are easily detected by use of pH indicator dyes, alcohols are more difficult to detect, and gases may be detected in inverted vials or by producing bubbles or cracks in agar deeps.

Disaccharides

Sucrose ⟶ glucose + fructose
Maltose ⟶ 2 glucose units
Lactose ⟶ glucose + galactose

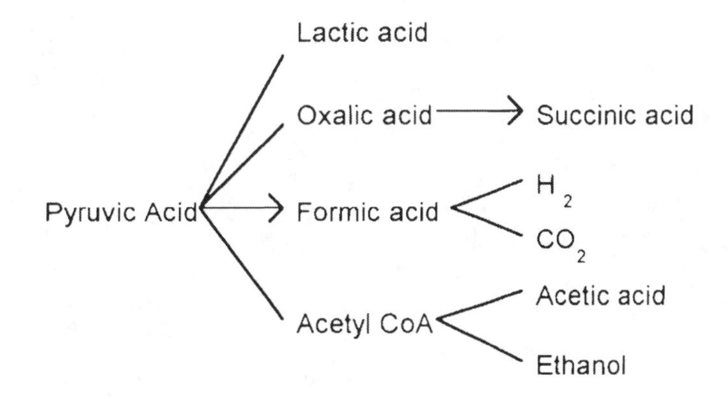

Procedure:

1. Durham tubes of glucose, lactose and sucrose broth will be provided. The broths contain brom cresol purple indicator [purple at pH 6.8 and yellow pH <5.2] in addition to other ingredients.

2. Inoculate one tube of each sugar with *E. coli*. Inoculate a second set with *Staphylococcus epidermidis* and a third set with your unknown.

3. Incubate all tubes at 30°C until next period.

4. Record in tabular form the production of acid and gas by the three sets of tubes.

Questions:

1. At what pH does brom cresol purple change color? What is its acid color? What is its alkaline color?

2. Might the results have been different if brom thymol blue had been used instead of brom cresol purple? Explain.

3. What fundamental physiological characteristics of bacteria may be learned by fermentation tests with various carbohydrates?

4. Do bacteria usually produce gas without acid from carbohydrates?

5. What benefit do bacteria derive from fermentation of sugar?

Indicator Dye	pH Range	Acid Color	Alkali Color
brom phenol blue	3.0 - 4.9	yellow	blue
brom cresol green	3.8 - 5.4	yellow	blue
brom cresol purple	5.2 - 6.8	yellow	purple
brom thymol blue	6.0 - 7.6	yellow	blue
phenol red	6.8 - 8.4	yellow	red
methyl red	4.4 - 6.2	red	yellow

EXPERIMENT 12

REDUCTION OF NITRATES

Many bacteria can utilize nitrates as hydrogen acceptors and reduce them to nitrites. Some can reduce the nitrites to ammonia or free nitrogen. The nitrates or nitrites may thus permit anaerobic growth of the bacteria. Most bacteria that utilize nitrates do so with an inducible system, which requires anaerobic conditions for induction. This form of nitrate reduction is called dissimilatory

$$\underset{\text{Nitrate}}{NO_3^-} + 2\,e^- + 2\,H^+ \longrightarrow \underset{\text{Nitrite}}{NO_2^-} + H_2O$$

$$2\,NO_3^- + 10\,e^- + 12\,H^+ \longrightarrow N_2 + 6\,H_2O$$

Procedure:

1. Inoculate a nitrate stab tube (semi-solid) with *Proteus vulgaris,* another with *Pseudomonas aeruginosa*. Use the inoculation needle and make one stab to the bottom of the tube.

2. Incubate the tubes at 30°C for 5 to 7 days.

3. Observe whether there is a bubble of gas in the agar. Gas would indicate that nitrates were used and converted to N_2, N_2O or NO_2.

4. Test for nitrate by adding to the culture 1 ml of sulfanilic acid reagent, followed by 1 ml of dimethyl-alpha-naphthylamine solution. The presence of nitrite is indicated by a pink or red color.

5. If the test is negative, test for unused nitrite by adding powered zinc. The zinc will convert the nitrate to nitrite which in turn will turn red in the presence of nitrate reagents. This would confirm the nitrates were not utilized by the bacteria.

6. Tabulate your results.

Caution: Dimethyl-alpha-naphthylamine: Avoid contact with skin.

Questions:

1. How does reduction of nitrate benefit the cell?

2. Under what conditions are nitrates reduced?

3. What becomes of the oxygen lost by nitrates or nitrites when they are reduced?

4. What is the source of energy necessary to carry out these reactions?

EXPERIMENT 13

PRODUCTION OF INDOLE

Indole is a putrefactive decomposition product formed by removing the side chain from the amino acid tryptophan. Since not all bacteria can bring about this reaction, it is useful as an aid in classification.

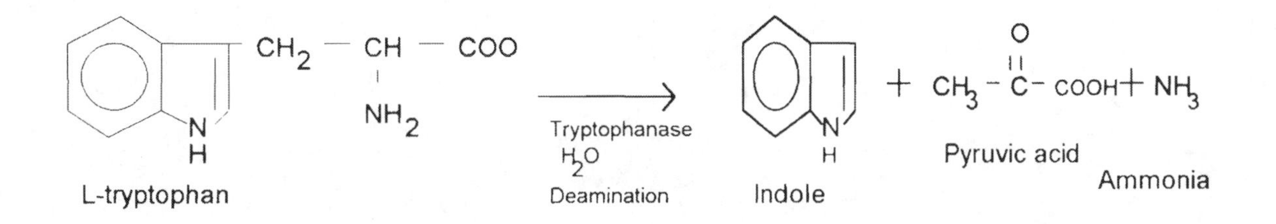

Procedure:

1. Tubes of tryptone broth (a digest of casein which is especially rich in tryptophan) will be provided.

2. Inoculate one tube with *E. coli,* another with *Enterobacter aerogenes.*

3. Incubate the tubes at 30°C until next period.

4. Test for indole by adding 10 drops of Kovac's reagent (a solution of p-dimethylamino-benzaldehyde in HCl with amyl alcohol) and shaking the tube slightly. A red layer on the top is a positive test.

5. Record the results.

Questions:

1. What is the formula for tryptophan? For indole?

2. Does any other naturally occurring amino acid contain the indole ring?

3. In what class of natural substances would one look for tryptophan?

4. Where in nature might one expect to find indole?

EXPERIMENT 14

PRODUCTION OF HYDROGEN SULFIDE

Cystine and certain other sulfur-containing amino acids are dissimilated by some bacteria with liberation of H_2S. This gas may be detected by its odor, but in microbiological tests the hydrogen sulfide reacts with heavy metals in the medium producing a blackening of the agar.

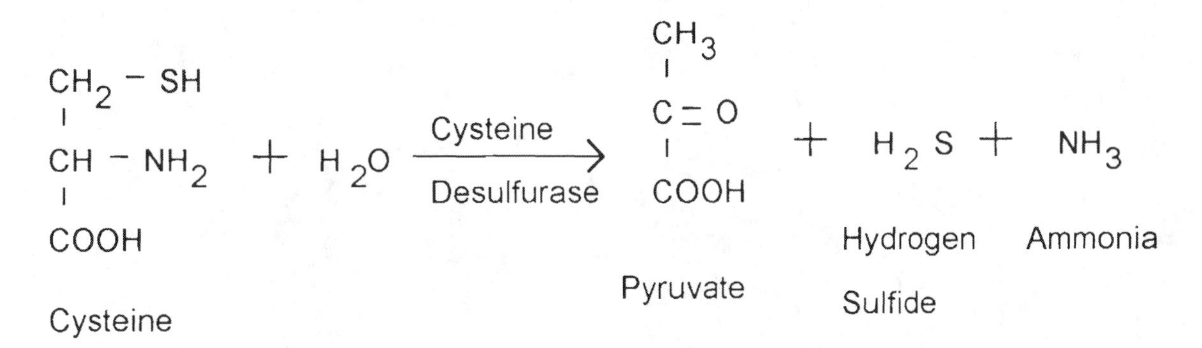

Procedure:

1. Inoculate tubes of peptone-iron agar by stabbing (using the inoculating needle) with *Proteus vulgaris* and any unknown you have.
2. Incubate the tubes at 30°C until next period.
3. Examine the cultures for blackening along the line of growth and out into the medium.
4. Record your results.

Questions:

1. What is cysteine? What is cystine?
2. How may hydrogen sulfide be formed from these compounds?
3. What is the reaction that resulted in blackening of the agar?
4. From what other type of compound may H_2S be formed from bacteria?

EXPERIMENT 15

REDUCTION OF HYDROGEN PEROXIDE

Catalase is an enzyme produced by many bacteria and is present in the greatest amounts in strictly aerobic bacteria, but is not produced by most obligate anaerobes. It is capable of decomposing hydrogen peroxide to water and molecular oxygen. The function of catalase is to remove the highly toxic product of aerobic metabolism, hydrogen peroxide. The genera *Streptococcus* and *Enterococcus* can be differentiated from *Staphylococcus* by whether they have the enzyme catalase.

$$2\ H_2O_2 \xrightarrow[\text{Catalase}]{} 2\ H_2O + O_2\uparrow \quad \text{(bubbles)}$$

Hydrogen Peroxide — Water — Oxygen

Procedure: Method 1 Agar plate

1. Mark sectors on the bottom of an agar plate with a wax pencil.
2. Inoculate the center of each sector with a loopful of each of the following: *E. coli, Enterococcus faecalis* & *B. subtilis,* plus any unknowns you have. Incubate at 30° C for two days.
3. Add a drop of hydrogen peroxide directly on top of the growth in each sector and observe for bubbles (oxygen gas).

Method 2 Slide test

1. Smear a small amount of bacteria from a slant or a plate onto a clean microscope slide.
2. Add a drop of hydrogen peroxide to the smear and watch for the production of bubbles, indicating the release of molecular oxygen.

Questions:

1. How does catalase differ from peroxidase?

2. Why is a piece of living plant or animal tissue able to decompose hydrogen peroxide?

3. Is catalase an intracellular or extracellular enzyme?

4. What other enzyme is important for organisms growing in the presence of oxygen?

EXPERIMENT 16

THE MR-VP CITRATE TEST

There are species of gram-negative bacteria that closely resemble each other in morphological and lactose-fermenting properties. Therefore, various tests have been devised to distinguish these organisms from one another. Four of these tests are the indole, methyl red, Voges-Proskauer and citrate tests. They are useful in special cases for securing further information regarding the probable source of coilform bacteria in water. The production of indole was already determined in experiment 13. The butanediol pathway is used by *Enterobacter* but not by *Escherichia coli*. One of the end products of the fermentation pathway, butanediol is a neutral product. The assay really detects the intermediate, acetyl methyl carbinol. The methyl red test really detects the presence of acids that drop the pH below 5, which is something that *E. coli* does. The citrate test determines whether an organism can utilize citrate as its sole source of carbon and energy.

Procedure:

1. Inoculate the following media with your unknown and control tubes with *E. coli* and *Enterobacter aerogenes*.

   a) Two tubes of MRVP medium (glucose-peptone phosphate broth)
   b) Citrate agar slant (streak on the surface of the agar).

2. Incubate the tubes at 30°C for 2-3 days (or until next period),

3. To one MRVP culture, add 0.6 ml of 5% alcoholic alpha-naphthol and 0.2 ml of 40% KOH. Vortex the tube vigorously and let it stand 15 minutes. A pink color denotes a positive test.

4. To the second MRVP culture, add 10 drops of methyl red indicator. A red color indicates that the reaction is below pH 5, and is recorded as a positive MR test. Yellow is a negative test.

5. Observe the citrate slant for growth or a blue color throughout the agar. Either is a positive citrate test. Lack of growth and a green color indicate a negative test.

6. Record your results.

7. Look around the class to see if others have different results from yours.

Questions:

1. For what substance is the VP reaction a test?

2. What is the reaction of the VP test?

3. What genus of coliform bacteria is characterized by production of acetyl methyl carbinol?

4. What is the physiological basis for the differentiation of *Escherichia coli* and *Enterobacter aerogenes* by the methyl red test?

5. Through what pH range does methyl red change color? What is its acid color? Its alkaline color?

6. What is a synthetic medium?

7. Why does growth of an organism on citrate slants indicate utilization of citrate?

8. Why is the blue (alkaline color) of brom thymol blue indicator produced when citrate is utilized?

9. What are the IMViC reactions of a typical *Escherichia*? Of a typical *Enterobacter*?

EXPERIMENT 17

CALIBRATION OF THE MICROSCOPE

The size of an organism is an important part of its description. Measurement of a microscopic organism requires some means of judging the extent of magnification provided by the microscope.

1. Observe the ocular micrometer in the eyepiece of your microscope. This bears a scale divided into fifty small divisions of arbitrary length. The apparent length of each division depends upon the magnification, so the scale must be calibrated for use with each combination of objective and eyepiece. To accomplish this a stage micrometer (a slide accurately ruled in units of 1mm., 0.1mm, and 0.01mm) is used. The calibration consists of determining how many divisions on the ocular micrometer correspond to a known distance on the stage micrometer.

2. Focus on the rulings of the stage micrometer with the low power objective. Move the stage micrometer so that any two of its markings coincide exactly with any two on the ocular micrometer. Record the distance on the stage micrometer and the corresponding number of ocular micrometer divisions in the table below.

3. Repeat step (2) with the high dry and oil immersion objectives.

4. Perform the required calculations to obtain the constants (c), which indicate for each objective the distance on a microscope slide represented by each ocular micrometer division.

Objective	Distance on stage (mm) micrometer x	Number of eyepiece divisions y	$c = \frac{x}{y}$ (mm)	c (μm)
Low Power				
High Dry				
Oil Immersion				

Stage micrometer number ________ Microscope Number ___________

EXPERIMENT 18

MEASURING BACTERIAL GROWTH - OPTICAL DENSITY METHOD

This experiment will be primarily an explanation and demonstration of the measurement of bacterial growth turbidimetrically using a spectrophotometer. The laboratory instructors will explain the use of the Spectronic 20 spectrophotometers; including an explanation of the relationship of turbidity (optical density) to numbers of bacteria.

In lecture, growth of bacteria was covered, including the parts of the growth curve. The calculation of the generation time was not covered. You will receive the O.D. data points from a growth experiment and plot the data on semi-log graph paper, during a quiz that you will have next laboratory meeting. The laboratory instructors will demonstrate how to draw a best-fit curve and how to extrapolate the generation time in minutes from the linear portion of the growth curve.

The experiment will run as follows: six tubes of pre reduced Trypticase broth will be inoculated with identical aliquots of *Clostridium difficile* at 6 A.M. The tubes will be incubated in the following manner.

Tube #1: Trypticase yeast extract - pH 6.5
Tube #2: Trypticase yeast extract - pH 8.5
Tube #3: Trypticase - pH 6.5
Tube #4: Trypticase - pH 8.5
Tube #5: Trypticase yeast extract glucose - pH 6.5
Tube #6: Trypticase yeast extract glucose - pH 8.5

There will be uninoculated tubes of media for zeroing the spectrophotometers.

You will be asked to compare the effects that the addition of glucose and the addition of yeast extract had on the growth of the bacterium, as well as the effect that the different pH had on these cultures.

Comparisons

pH: tube #1 vs. #2; #3 vs. #4 and #5 vs. #6

glucose: tube # 1 vs. #5 and #2 vs. #6

yeast extract: tube #1 vs. #3 and #2 vs. #4

EXPERIMENT 19

ENUMERATION OF LIVE BACTERIA

STANDARD PLATE COUNT

If bacteria are evenly dispersed in a broth culture and a sample is spread over an agar plate so that each bacterium forms a separate colony, then the number of live bacteria in a culture can be accurately estimated. Sometimes the bacteria are in pairs or clumps so that colonies formed do not represent growth from a single bacterium. When estimating the number of bacteria in a sample, these colonies are refered to as "colony forming units" (CFU).

A standard plate count method has been established for water and milk analysis. Liquid samples containing bacteria are added to an empty Petri plate. Melted agar is poured into the plate and then mixed in with the samples.

Procedure:

1. Shake the sample in the bottle vigorously, 25 times and in a 45° arc.

2. Dilute out the sample to provide 25 to 250 colonies per one ml of the test sample. To do this use a series of three "blank" test tubes, each containing 9 ml of peptone water and make 1 ml serial transfers. Use a new pipette for each dilution. The first tube will be a one to ten dilution (1:10), the second a 1:100 and the third a 1:1,000 dilution. Mix the tubes thoroughly.

3. Upon making the second dilution (1:100), use a fresh pipette and put 1 ml of the 1:10 dilution in the second tube and one ml in the center of each of two empty Petri dishes. Do the same for the 1:1,000 dilution. Mix each tube thoroughly when making dilutions.

4. After making the 1:1,000 dilution use a new pipette and pipette 1 ml to each of two Petri plates and 0.1 ml (two drops) to two other plates.

5. Label all plates as to the dilution, source of sample and your name.

6. When you are completely ready, add agar to one plate at a time. Add the agar until the bottom is covered, then mix the sample before adding agar to the next plate. Agar hardens at 42° C so do not waste time during the pouring process.

7. To mix samples make three moderately vigorous swirls to the right and three to the left.

8. Incubate the plates at 30° C.

9. At the next laboratory session, select plates that have between 25 and 250 colonies. Average the numbers on countable plates and report results in two significant figures.

Example: [Those plates with counts 126, 136, and 111 would be reported as a plate count of 120. If the dilution was 1:1,000, then the count would be 120,000 or 1.2×10^5 CFUs per ml of original sample.]

EXPERIMENT 20

BACTERIAL MUTATION: THE ISOLATION OF A STREPTOMYCIN-RESISTANT MUTANT

Because of the low rate of appearance of mutant cells and because special methods must be used to select out these mutants, the demonstration of a mutant cell is not an easy task. Changes in the physical or chemical environment will accelerate the frequency of appearance of these mutants. For example radiation with x-rays or ultraviolet light will increase the rate of mutation. If the mutant cell, by virtue of the change which it has undergone, is somehow especially suited to develop in the environment in which it is formed, it may outgrow its parent culture and become dominant.

Spontaneous mutants resistant to an antibiotic are readily detected since they can grow in the presence of the antibiotic which inhibits growth of the wild type. When the bacteria are grown on media containing the specific antibiotic, this is called selective pressure, because there is a selective advantage for the mutants to grow.

Procedure:

1. Nutrient agar plates containing: 0.01, 0.1, 0.5, and 1 mg/ml streptomycin will be provided.

2. Pipette approximately 4 drops of *Serratia marcescens* onto each agar plate. Using a sterile glass spreader (sterilized by dipping in 95% alcohol, flaming and cooling it on the sterile surface of the agar), spread the culture over the agar surface.

3. Do a plate count on the diluted sample (10^{-6}) of the *S. marcescens* on nutrient agar containing no streptomycin (this is the control for determining the actual number of bacteria plated). Incubate all the plates at 30°C for four to seven days.

4. Count the number of colonies (mutants) on the streptomycin plates and count the number of colonies on the plate containing no streptomycin.

5. Estimate the frequency of mutants [the ratio of mutants to the total number of cells plated (the number of colonies on the control nutrient agar plate times the dilution factor)].

Alternate Procedure:

1. Nutrient agar plates containing: 0 and 100 μg per ml streptomycin will be supplied.

2. Pipette 2 drops (0.1 ml) of *Escherichia coli* K-12 (wild-type) onto a nutrient agar plate containing streptomycin. Using a sterile glass spreader (sterilized by dipping in 95% ethanol and lighting in a flame) spread the inoculum evenly over the agar surface.

3. Do a plate count on the diluted sample (10^{-6}) of the *E. coli* on nutrient agar containing no streptomycin (this is the control for determining the actual number of bacteria plated). Incubate all the plates at 30°C for four to seven days.

4. Count the number of colonies (mutants) on the streptomycin plates and count the number of colonies on the plate containing no streptomycin.

5. Estimate the frequency of mutants [the ratio of mutants to the total number of cells plated (the number of colonies on the control nutrient agar plate times the dilution factor)].

EXPERIMENT 21

BACTERIAL VARIATION

The microbial cell may be regarded as a complex of morphological structures and interlacing chemical activities, which under certain conditions can exhibit a response to the environment. Thus, capsules of *Alcaligenes viscolactis* are produced at low temperatures but not at high temperatures, although the organism grows successfully at the higher temperature. Flagellation of certain bacteria is more pronounced at room temperature than at 37°C, and some organisms produce a pellicle aerobically, but not anaerobically. Nutritional variants are also known. *Lactobacillus casei* can be grown at pH 5 without adding the B-vitamin "pyridoxine" to the medium, but at pH 7 pyridoxine must be added.

This exercise illustrates a variation in the pigmentation of *Serratia marcescens* in response to different temperatures. Interpret your results in terms of existing knowledge of bacterial physiology and genetics.

Procedure:

1. Prepare two nutrient agar streak plates of *Serratia marcescens*.

2. Incubate one at room temperature and the other at 37°C until the next laboratory period.

3. Observe the effects of the growth temperature on the production of pigment by *Serratia marcescens*.

Questions:

1. What is enzymic induction?

2. Does phenotypic variation without genotypic variation occur in higher forms of life?

3. Design an experiment to show that the color change was the result of enzyme induction rather than a mutation.

4. How would you show that the color change was the result of enzyme inactivation rather than induction or mutation?

EXPERIMENT 22

CULTIVATION OF ANAEROBIC BACTERIA

For the cultivation of most strictly anaerobic bacteria, oxygen must be removed and the oxidation-reduction potential must be lowered. Autoclaving the media expels oxygen from the media and trace amounts of remaining oxidized molecules are scavenged up by addition of thioamino acids (reducing compounds). Usually, a reducing agent like thioglycollate is added to the medium to ensure that there is sufficient reducing power available. A small amount of agar is frequently added to broth media to help exclude oxygen. The medium usually contains an oxidation-reduction indicator dye (resazurin or methylene blue) that will be colored in the presence of oxidized molecules. The other method for the cultivation of anaerobic bacteria is the use of gas generation packages in special jars (Gas Pak @), for the production of colonies on agar plates.

Procedure: A. Removal of Oxygen - the Gas Pak jar

1. Work with one jar for two bench sides. Inoculate two sets of slants with *Clostridium sporogenes, E. coli*, and *Pseudomonas aeruginosa*.

2. Put one set of slants in the Gas Pak jar, open an indicator strip and put it in the jar, open a gas-generating envelope, add 10 ml of water to the envelope and add it to the jar. Seal the jar and incubate it at 30°C until next period. Put the other set of tubes in your incubator bin at 30°C.

3. After 4 days remove the slants from the jar and observe for growth. Smell each culture through the cotton plug and correlate these smells with other experiments.

B. Removal of oxygen by reducing agents

1. Inoculate a tube of freshly boiled thioglycollate semi-solid agar medium with one loopful of *C. sporogenes*.

2. Inoculate a tube of nutrient broth with the same organism. Incubate both tubes at 30°C until next period.

3. Record presence or absence of growth in the two tubes.

4. Prepare a gram stain from the thioglycollate culture and observe the gram reaction and the shape of the sporangium.

5. Sketch a few cells and record the gram reaction.

Procedure:

1. Inoculate a tube of nutrient broth with a loopful of *Enterobacter aerogenes*.
2. Inoculate a second tube of nutrient broth with a loopful of *C. sporogenes*.
3. Inoculate a third tube of nutrient broth with a loopful of each organism.
4. Incubate the tubes at 37°C until next period.
5. Observe the tube for character of growth and odor.
6. Prepare gram stains from each culture showing turbidity.
7. Record your results.

Questions:

1. What is one theory to account for the inability of anaerobes to grow in the presence of free oxygen?

2. Why is a trace of agar put into thioglycollate media?

EXPERIMENT 23

AGGLUTINATION OF BACTERIA

When bacteria in saline suspension are mixed with an antiserum which has been prepared by immunizing an animal with the same organism (or which has been taken from an immune person), the bacteria are clumped together. This process is called agglutination. Agglutination is highly specific; that is a serum which agglutinates one species of organism may not agglutinate any other. This is therefore a useful test for identification of unknown bacteria.

Procedure:

1. Mark two rings the size of a nickel on a glass slide with a wax pencil.
2. Place in each ring a drop of the heavy merthiolate-killed bacterial suspension provided.
3. To one ring add one drop of diluted antiserum against the same bacterial species; to the other add a drop of saline.
4. Mix the drops with a sterilized needle (flame between drops) and then rock and tilt the slide a few moments.
5. Observe granulation and clumping in the mixture containing antiserum. The other (control) mixture should not agglutinate.

Questions:

1. What is an antigen? What is the test antigen in this experiment?
2. What are antibodies? Where are they found?
3. What other reactions besides agglutination may be demonstrated between antigens and antibodies?
4. Indicate two ways agglutination can be used to help diagnose infectious disease?

EXPERIMENT 24

IMMUNE PRECIPITATION
(THE GEL DIFFUSION TEST)

When a soluble antigen (e.g. , a protein) and its homologous antibody are mixed in optimal ratio, precipitation occurs due to formation of a three-dimensional lattice formed by multivalent antigens reacting with bivalent antibodies. Establishment of the necessary optimal ratio is facilitated by allowing the reagents to diffuse toward each other through the agar gel. The resulting precipitate has the form of a line or an arc.

An antiprotein antiserum and its homologous protein antigen will be provided, together with two other protein solutions labeled A and B. Prepare a film of 1% ion agar (in borate-buffered saline, pH 8.5) on a clean microscope slide by pipeting 2.5 ml of melted agar to the slide in such a manner as to cover the entire slide uniformly. Set aside to harden.

When the agar has hardened thoroughly, remove plugs of agar by use of a bent Pasteur pipet attached to an aspirator using a template such as the following:

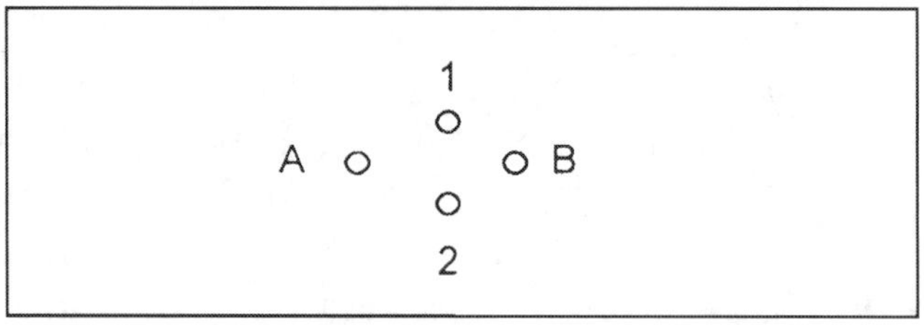

```
        1
        o
A o           o B
        o
        2
```

Take antiserum into one end of a hematocrit or melting point capillary tube by capillary, and then touch well no. 1 with the end of the tube to permit the well to fill by capillary. Break off and discard the end of the capillary tube. Take protein homologous to the antiserum into the other end of the tube and fill well no. 2. Discard the remainder of the tube.

With a fresh capillary tube place proteins A and B in the other two wells. Keep the slide at room temperature in a moist chamber. Lines of precipitate may form as early as 2-3 hours with homologous reagents and usually remain distinct for 2 or 3 days. Determine whether A and/or B is the same protein as no. 2.

EXPERIMENT 25

SELECTIVE BACTERIOSTATIC ACTION OF CRYSTAL VIOLET

Growth of certain bacteria can be inhibited by triphenylmethane dyes although growth of other bacteria may occur. This phenomenon is known as selective bacteriostasis.

Procedure: 1. Nutrient agar plates containing 0.001 g/L crystal violet will be provided.

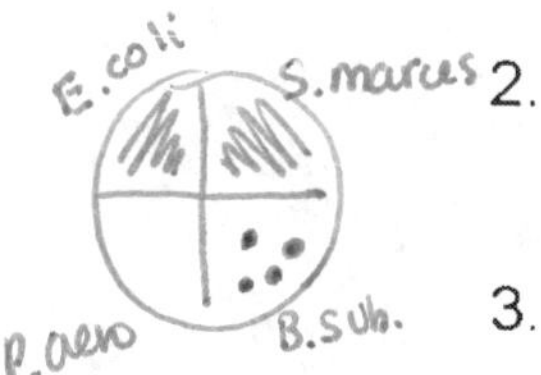

2. Mark the plates into quarters and label as follows: (a) *E. coli*; (b) *Serratia marcescens*; (c) *B. subtilis*; (d) *S. epidermidis*.

3. Inoculate each section by making a short streak with a loopful of the appropriate broth cultures.

4. Incubate the plate at 30°C until next period.

5, After incubation, record the relative amount of growth of the various bacteria.

6. Sub culture an inoculum from where no bacteria grew to a nutrient agar plate and incubate the plate at 30 °C until next period.

Questions: 1. What are the Gram reactions of these bacteria?

2. Is there any correlation between Gram reaction and inhibition by crystal violet? Can you offer an explanation?

3. Why are dyes referred to as bacteriostatic agents?

4. Do all Gram positive organisms exhibit the same degree of sensitiveness to bacteriostatic dyes?

5. Describe a practical application of selective bacteriostatic action in the purification of mixed cultures.

6. Could crystal violet be used in the treatment of infections, as far as this experiment indicates?

EXPERIMENT 26

OLIGODYNAMIC ACTION OF METALS AND BACTERIOSTATIC ACTION OF CHEMICALS

An easy, but only approximate, method for comparing bacteriostatic agents consists of testing their ability to inhibit growth of a test organism on an agar plate when the agents are placed on the surface of the agar. Antibiotic susceptibility assays such as penicillin assays are often done by this method.

Procedure:

1. Divide the bottom of a Petri dish into four sections with a glass marking pencil. Place a clean silver or copper coin in the center of one section, after you have dipped it in alcohol and flamed it.
2. Inoculate a tube of melted and cooled nutrient agar with several loopfuls of *S. epidermidis.* Mix and pour into the Petri dish. Allow the agar to harden.
3. In the center of another section of the plate place a small bit of an ointment provided (a piece of filter paper).
4. In the other sections place filter paper disks which have been soaked in household disinfectants or antiseptics.
5. Incubate the plate, inverted as usual, at 30°C until the next lab period.
6. After incubation, measure the width of the clear zone around each substance. Also sketch the appearance of the plate.
7. Subculture, to a nutrient agar plate, from each of the sectors where no bacteria grew.

Roccal & Ethanol have least growth
Iodine has most growth
Penny in the middle

Iodine has no restrictive properties

Questions:

1. Would a completely insoluble substance exert any antiseptic action in this test?

2. What does oligodynamic mean? How is the term appropriate as used in this experiment?

3. Does the phenomenon of oligodynamic action have any practical application?

4. How is the width of a clear zone around an antiseptic related to the concentration of the chemical?

5. How would you demonstrate that the chemical may have been bactericidal rather than bacteriostatic?

more growth of bacillus subtilis
less growth of E.coli
(E.coli impeded less)
streaked down the middle

EXPERIMENT 27

ANTIBIOSIS

Antibiosis has received much publicity, due to the development of penicillin, streptomycin and other similar substances. It has been defined as the harmful effect of one microorganism on another. Waksman defines an antibiotic as a specific substance produced by a microorganism which exerts an inhibitory effect on certain other organisms. Bacteria of the genus *Streptomyces* are very common to the soil and they are among the major producers of antibiotics that we use for treating microbial infections. In this experiment, we will isolate some *Streptomyces* from soil and assay them for antibiotic activity against several bacteria. We will employ a selective agent (cycloheximide 50 μg/ml) in the medium used to isolate the *Streptomyces* from soil. The selective agent is used to specifically inhibit the growth of Eucaryotic organisms, molds, since soil is heavily laced with mold spores.

Procedure:

Part I

1. If not already provided, pour a plate of yeast glucose agar or tryptose agar and allow it to harden. Inoculate it with a single streak of *Penicillium notatum*, toward one side of the plate. Incubate at 30°C for 2 - 5 days.

2. When there is good growth of the mold, inoculate the plate by making streaks at right angles to the mold growth with: *E. coli, M. luteus* and *B. subtilis*.

3. Incubate the plate until next period at 30°C.

4. Measure the zone of inhibition of the test organisms.

Part II

1. Make a pour plate of soil bacteria by placing 1 ml of a soil suspension in the center of an empty Petri plate and then mix in agar. Incubate the suspension at room temperature until the next lab period.

2. Observe the plate: looking for colonies with clear zones around them, where no other bacteria grew.

Part III

1. Suspend a gram of rich soil in several milliliters of water in a test tube. Select a plate of Emerson agar and label it in the prescribed manner.

2. Obtain a loopful of the soil suspension and streak it for isolated colonies as noted in exp. 3. Incubate the plate for one week at room temperature.

3. Examine the plate for hard white colonies shaped like tiny volcanoes. The plate should have a pungent, earthy odor. The material in the colony will be dry and hard to remove, but a section should be pried from the edge of the colony and examined in a wet mount preparation under the high power objective (remember to use the iris diaphragm to decrease the amount of light to see the specimen in the wet mount). The organism should have conidia.

4. Obtain another plate of Emerson agar and label it. Using a loop, take a section of the *Streptomyces* colony and make a single streak across the plate. Incubate the plate at room temperature for 3 or 4 days, until growth is apparent.

5. After the initial incubation period, make single streaks of available test microorganisms at right angles to the *Streptomyces* streak, use a sterile swab for each of the different organisms you inoculate. Be careful not to touch the *Streptomyces* growth with the swab. Re incubate the plate at room temperature for another week.

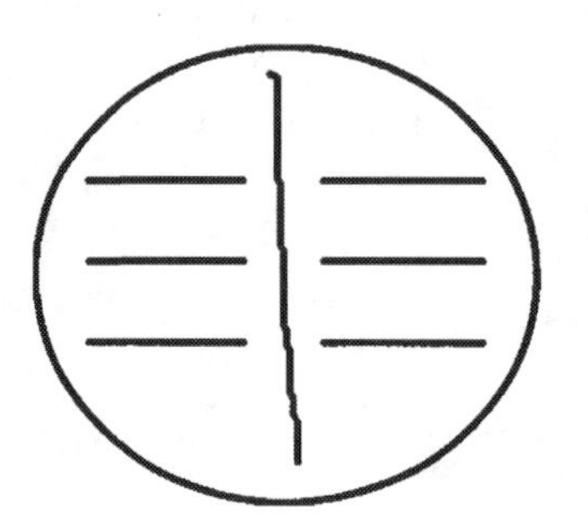

6. Examine the growth of the test organisms where the streak crosses the *Streptomyces*. Note a reduction in the amount of growth of the test organism, indicative if antibiotic activity. Illustrate your results and draw your conclusions from your data.

Questions:

1. What is the substance that inhibits one of the test organisms of this experiment?

2. Is this substance excreted into the culture medium? What evidence is there?

3. Could you demonstrate this antagonistic effect by use of sterile culture filtrates of molds or other organisms?

EXPERIMENT 28

MORPHOLOGY OF YEAST

Yeasts are important microorganisms, both from the point of view of the spoilage which they cause and the uses to which they may be put. Yeasts belong to the Fungi Kingdom and are Eucaryotic cells. They are considerably larger than most bacteria, and are easily studied under the microscope.

procedure: Young Actively Growing Yeast Culture

1. Prepare a wet mount of *Saccharomyces cerevisiae* by placing several loopfuls of the 6-8 hour yeast glucose broth culture provided on a microscope slide; then lay a coverslip over to yeast. (Vortex the tube, before preparing the slide)

2. Observe the cells with the high dry (43X) objective. Note evidence of budding. Sketch the largest aggregation of cells you find and indicate if possible, which you think was the original mother cell.

3. Prepare a smear from the broth culture (do not add water to the slide), let it dry heat-fix as usual and Gram stain.

4. Observe under oil and observe the Gram reaction of the cells. How does their size compare with that of the bacteria you have seen? (Use your calibrated optical micrometer to measure the cells)

5. Measure the length and width of the cells. Compare the size with that of bacteria: *E. coli and B. subtilis*.

Questions:

1. Why is a young culture of yeast used in the first part of this experiment?

2. Do yeast have any other method of reproduction than budding?

3. Does *S. cerevisiae* produce spores?

4. What names are sometimes given to yeasts that do not produce spores?

5. Do yeasts contain a nucleus that can be seen in wet mounts or Gram-stained smears?

6. When are granules especially prominent in yeast cells?

7. Are yeasts motile?

8. How are yeasts differentiated from bacteria?

9. What is the chemical composition of the yeast cell wall?

10. What is the equation for the alcoholic fermentation?

EXPERIMENT 29

MORPHOLOGY OF SOME COMMON MOLDS

Molds are interesting microorganisms in their own right, but are so widely distributed in Nature that they are also of concern in a bacteriology laboratory as contaminants in cultures of other organisms. Moreover, industrial spoilage problems involving molds often come to the attention of microbiologists.

Procedure:

1. Demonstration plate cultures of *Rhizopus, Aspergillus, and Penicillium* will be available. Tabulate a description of these under the following headings:

   a) Diameter of colony in millimeters
   b) Color of colony
   c) General appearance [cottony, dusty, velvety, etc.]

2. Sterilize a hanging drop slide by dipping it in alcohol and burning it off. Sterilize a small coverslip in the same manner. Be careful not to heat it too much or it might break.

3. Water suspensions of spores from the three molds will be available. Select one which your two neighbors do not use.

4. Secure a tube containing melted and cooled (45-50°C) yeast glucose agar from the water bath. **Immediately inoculate** it with a loopful of the spore suspension. **Mix thoroughly** by rotating the tube between the palms of your hands.

5. Transfer to one edge of the depression in the slide enough of the seeded agar to fill about one-third of the depression. **Before the agar hardens**, place the coverslip over the depression so that the agar spreads around the depression. Leave a tiny space for air at the opposite side of the depression.

6. Place the slide in a Petri dish containing a moistened piece of filter paper. Incubate it at 30°C until the next period. [incubate it at room temperature, if it will go over a weekend].

7. After incubation, observe your preparation with the low power and high dry objectives of the microscope. Reduce the light with the iris diaphragm, if necessary , and be careful not to focus downward.

8. Sketch a portion of the mold showing a typical fruiting body. Label structures. Note and label the following:

   a) Fruiting bodies (shape, color?)
   b) Vegetative mycelium (septate or nonseptate?)
   c) Fertile hyphae (septate or nonseptate?)

9. Repeat with your neighbors' preparations.

Questions:

1. How do molds differ from yeasts? From bacteria?

2. What is the composition of the cell walls of molds?

3. Are mold spores more resistant to unfavorable conditions than bacterial spores?

4. Distinguish between sporangiospores and conidia.

EXPERIMENT 30

EFFECT OF THE ENVIRONMENT UPON MICROORGANISMS

For each organism there is a minimum growth temperature, or temperature below which growth will not occur; an optimum growth temperature, at which growth occurs most rapidly or in greatest amount; and a maximum growth, above which growth will not take place. This experiment demonstrates differences among microorganisms.

Procedure:

1. Use one of the following: *Enterococcus faecalis, Bacillus subtilis, Escherichia coli, Bacillus stearothermophilus, Pseudomonas aeruginosa* and *Saccharomyces cerevisiae*.

2. With the organism selected, inoculate uniformly, by loop, four yeast glucose agar slants or other media as directed.

3. Mark the tubes for incubation at 10°C, 20°C, 37°C and 55°C, respectively, and place them in racks designated for incubation at these temperatures.

4. After several days compare the amount of growth on the various slants and grade them accordingly:

   - = no growth
   + = poor growth
   ++ = fair growth
   +++ = good growth

5. Tabulate in your notes your own results and those of your neighbors so that you have data on all the organisms.

Questions:

1. Can you think of an explanation for the fact that not all bacteria have the same optimum growth temperature.

2. What is the meaning of psychrophilic? Mesophilic? Thermophilic?

3. Discuss the mechanism of growth control by temperature.

4. Why do organisms fail to multiply when kept at temperatures above the maximum and below their minimum.

EXPERIMENT 31

EFFECT OF OSMOTIC PRESSURE ON MICROORGANISMS

Various microorganisms are inhibited or killed by different concentration of solutes, such as NaCl, as a result of their effect on the osmotic pressure of the medium. The comparative effect on several bacterial species, a mold and a yeast will be demonstrated.

Procedure:

1. Tubes of yeast glucose broth containing NaCl in concentrations of 0.5, 3, 7, and 15% respectively will be provided. Normal broth contains 0.5% NaCl. Inoculate tubes with *S. cerevisiae, E. faecalis, E. coli, B. subtilis, B. stearothermophilus* or *P. aeruginosa*.

2. Inoculate the salt broth tubes with one loopful each of the organism you used in Exp. 32. Incubate as previously directed [remember *B. stearothermophilus* only grows at 55°C].

3. After incubation record the relative amounts of growth in each tube (-, +, ++, +++).

4. Secure a plate of yeast glucose agar. Mark the bottom of the dish into four sectors. Label them for each salt concentration, and inoculate each section with a loopful from the corresponding tube. Incubate at the proper temperature.

5. After incubation of this subculture, record the relative amount of growth in the various sectors. Tabulate your results on the next page.

Questions:

1. What is plasmolysis?

2. Is plasmolysis necessarily lethal to microorganisms?

3. What is a hypertonic solution?

4. Give some commercial applications of this phenomenon.

NaCl Concentration	0.5%		3%		7%		15%	
Organism	broth	agar	broth	agar	broth	agar	broth	agar

EXPERIMENT 32

EFFECT OF pH ON GROWTH OF MICROORGANISMS

The hydrogen ion concentration of the medium exerts a profound influence on the growth of microorganisms. For each organism their is an optimum hydrogen ion concentration at which growth occurs most rapidly or most abundantly. Their are also minimum and maximum values below and above which growth will not occur.

Procedure:

1. Tubes of yeast glucose broth buffered at pH 3, pH 5, pH 7 and pH 9 will be provided.

2. Inoculate one set of these tubes with a loopful each of the same organism used in Ex. 32. Incubate as previously directed.

3. After incubation record the relative amounts of growth (-, +, ++, +++) in each tube.

4. Tabulate your own data and those of your neighbors with the other organisms.

Organism	pH 3	pH 5	pH 7	pH 9

Questions:

1. Can you suggest any reason why certain microorganisms grow better in solutions of one pH than another.

2. How do molds and yeasts differ from bacteria with respect to their pH requirements? Does this have any practical application in food preservation?

3. Explain the action of buffers.

4. What is pH?

5. How might you determine the pH of a solution?

6. At what reaction (pH) are culture media for bacteria usually adjusted? for molds?

7. How does one adjust the pH of a culture medium?

EXPERIMENT 33

EFFECT OF FREEZING ON MICROORGANISMS

Freezing is an important method of preserving food, biologicals, and many other materials. It is therefore of interest to know whether this process kills microorganisms.

Procedure:

1. Inoculate a tube of yeast glucose broth with a loopful of the same organism used in Exp. 35. Mix the contents of the tube by vortexing it.
2. Place the tube in a rack designated to be frozen.
3. At the next laboratory period, thaw the tube at room temperature and incubate at the proper temperature.
4. Tabulate your results and those of your neighbors.

Questions:

1. Would you expect frozen foods to be sterile?

2. Explain the difference between the effect on microorganisms of very high temperatures and very low temperatures.

EXPERIMENT 34

FLORA OF THE SKIN

The skin flora consists of "normal" inhabitants plus adventitious organisms. Any of these may on occasion produce disease; i.e., if a wound permits them to get into the underlying tissues, if the physical resistance is lowered or if they find their way to another (susceptible) individual. The skin sampled may be that of the washed hands, the armpits, between the toes, or any other surface including the external auditory canal. Since these bacteria may be potential pathogens, use proper techniques and notify the instructor of spills. Discard plates in Biohazard containers and wash your hands before leaving the laboratory.

Procedure:

1st day: Carefully collect a specimen on a swab (if you are swabbing the face or the hands; moisten the swab in sterile saline, express the excess saline from the swab by pressing it against the inner wall of the tube). Roll the swab over one fourth of a blood agar plate, then over one-fourth of a phenethyl alcohol agar plate. Streak (with your loop) the two plates for isolation of colonies [the phenethyl alcohol agar plate is selective for staphylococci and fecal streptococci and may be streaked more heavily, since it contains a selective agent]. Label the plates and give them to the instructor for incubation at 37°C . {Your laboratory instructors will inoculate a culture of *Staphylococcus aureus*}

2nd day: After incubation, describe and Gram stain representative colonies on the two plates. Pick a staphylococcal colony and streak it (with your loop) for isolation on a mannitol salts agar plate. If you find a Gram-negative rod-shaped bacterium, streak it for isolated colonies on an EMB agar plate. If you find a Gram-positive non-sporing rod in palisade (parallel) arrangement, streak it for isolation on a blood agar plate. Incubate these plates at 37°C for one or two days. {Your laboratory instructor will inoculate and EMB plate with *E. coli*, a blood agar plate with a *Corynebacterium* and a mannitol-salts agar plate with *S. aureus* for demonstrations}.

3rd day (a) **Mannitol salts agar**. Ninety-five percent of *Staphylococcus aureus* produce a yellow (acid) halo around the bacterial growth on this medium. Confirm that your isolate is *S. aureus* by performing a coagulase test. Make a heavy suspension of the organism with a large loopful of water on a microscope slide, add a drop of rabbit plasma and mix with a clean toothpick. Rock the slide back and forth for one minute. A positive coagulase reaction consists of clumping the bacteria together (looking like cottage cheese).

(b) **EMB plate**. This medium inhibits most gram-positive and some gram-negative organisms (selective agents are Eosin Y and methylene blue dyes) and is differential (lactose and sucrose) for screening enteric gram-negative opportunists from pathogens. Carefully gram stain isolated colonies [dark colonies = fermenters (opportunists) vs. clear colonies = nonfermenters (pathogens)]. Pick (with an inoculating needle) from the center of the colony {the selective agents are inhibitory, but may not be bactericidal} and inoculate a triple sugar iron (TSI) agar slant by streaking the slant, stabbing the butt and re streaking the slant. Incubate these cultures at 37°C and observe at 24 and 48 hours.

(c) **Blood agar plate.** Corynebacteria usually produce mucoid, circular, raised colonies (generally non hemolytic). Gram stain isolated colonies. If you find gram-positive rods in a palisade (picket fence) arrangement; inoculate with a needle, by stabbing a set of CTA carbohydrates (glucose, maltose and sucrose) and inoculate a tube of urea broth. [Bacteria that hydrolyze urea - liberating ammonia produce an alkaline reaction that changes the color of the phenol red pH indicator from light yellow to red.] Incubate all plates and tubes at 37°C.

<u>4th day:</u>

(a) **TSI slant**. The medium contains 0.1% (wt/vol.) glucose, 1% (wt/vol.) sucrose, 1% (wt/vol.) lactose, phenol red (pH indicator), sodium thiosulfate (sulfur source for production of H_2S), and ferrous ammonium sulfate (Fe^{++} source for making FeS).

(1) If the organism can only ferment glucose, the whole tube turns yellow (indicating acid production). Then bacteria growing on the slant will oxidize the acid to CO_2 and H_2O and turn the slant red (alkaline) by an oxidative deamination of the amino acids in the peptone.

(2) If the organism ferments lactose (and/or sucrose), the whole tube will remain yellow (acid) because the concentration of these sugars allow for the production of large amounts of acid.

(3) If gas (CO_2 and/or H_2) is produced by the fermentation of glucose, there will be bubbles of breaks in the agar butt of the tube.

(4) If the organism produced H_2S from sodium thiosulfate, the butt of the tube will blacken (FeS). This automatically means acid was produced from glucose, since H_2S is only produced under acid conditions. (Organisms that do not ferment lactose are possible pathogens.)

Slant	Butt	Gas	H_2S	Possible Organisms (Genera)
K	A	-	-	*Shigella, Serratia, Proteus*
K	A	-	+	*Salmonella typhi*
K	A	+	-	*Klebsiella, Enterobacter, Proteus, Providencia*
K	A	+	+	*Proteus, Salmonella, Arizona, Citrobacter*

{K = alkaline (red); A = acid (yellow)}

(b) Read CTA carbohydrates yellow (acid) and urease red (alkaline) as positive reactions. Some healthy people carry *Corynebacterium diphtheriae*; however, most of these strains are avirulent.

Organism	Hemolysis	Glucose	Maltose	Sucrose	Urease
C. diphtheriae	+	+	+	-	-
C. pseudotuberculosis	+	+	+	±	±
C. xerosis	-	+	+	+	-
C. renale	-	+	-	-	(-)
C. pseudodiphtheriticum	-	-	-	-	+
C. haemolyticum	-	+	+	±	-
C. ulcerans	+	+	+	-	+

EXPERIMENT 35

FLORA OF THE UPPER RESPIRATORY TRACT

A variety of bacteria normally reside in the mouth, throat and nasal passages (nares). Some of them produce disease if they find their way to other less well protected parts of the body or to other individuals. The normal flora includes: staphylococci, streptococci, neisseria, diphtheroids, gram-negative rods and spirochetes. Since you are isolating pathogens, take proper care.

Procedure:

1st day: Swab your throat with a sterile swab. Inoculate a blood agar plate and a mannitol salts agar plate with the swab by rolling it over a quarter to a third of the plate, then discard the swab back into the tube it came in. Asepticly, with your loop, streak the two plates for isolated colonies. Label the plates and incubate them at 37°C.

Swab your throat with another sterile swab and make a smear, by rolling the swab on a microscope slide; gram stain the smear and examine it. Besides bacteria, you will also observe some large epithelial cells. Also gram stain the stock cultures of: streptococcus, neisseria and pneumococci.

Collect tartar (plaque) from between your teeth, especially your molars, with a tooth pick and prepare a smear using a loopful of nigrosin stain (diluted 1:5 with water). Let the smear air-dry and examine under oil. Look particularly for spirochetes and long tapering fusiform bacteria.

2nd day: Describe the colonies on the blood agar plate. Gram stain representative types. Viridans group streptococcal colonies are surrounded by a zone of α-hemolysis (greenish or brownish discoloration); β-hemolytic colonies are surrounded by a clear zone (complete lysis of the red blood cells), and γ-hemolytic colonies do not cause any change in the red blood cells.

Streak a viridans group streptococcus onto a fresh blood agar plate (streak it heavily over at least half of the plate). On the heavily streaked portion of the plate, deposit a differential optochin disk and sensitivity disks for penicillin and streptomycin. The α-hemolytic streptococci are hard to tell from the pneumococci; however, the pneumococci are sensitive to optochin, whereas the viridans streptococci are not. If a β-hemolytic streptococcus is found, streak it onto another blood agar plate and deposit a bacitracin differential disk and sensitivity disks for penicillin and streptomycin on the heavily inoculated area of the plate. Group A (human pathogenic) β-hemolytic streptococci are differentiated by bacitracin, and their growth will be inhibited in the immediate vicinity of the disk. Incubate the plate at 37°C.

Flood a portion of your original blood agar plate with oxidase reagent, pour off the excess, and observe a pink to purple color develop rapidly on some of the colonies. They may be gram-negative diplococci (Neisseria). The reagent will kill the bacteria, so quickly pick one such colony and streak for isolation on a blood agar plate. Incubate the plate at 37°C.

Observe the mannitol salt agar plate. Make gram stains from the staphylococcal colonies. If you find a mannitol positive colony, perform a rapid slide coagulase test. If the bacterium is coagulase positive, streak it onto a fresh blood agar plate to observe hemolysis and incubate it at 37°C.

3rd day: Observe the blood agar plates. If you have inoculated a blood plate from the mannitol salts agar plate observe for β-hemolysis (possibly for a double-zone).

Observe the blood agar plate that you plated an α-hemolytic streptococcus, especially for the reaction to the optochin disk (if growth is inhibited, the organism is *S. pneumoniae*).

Observe the blood agar plate that you transferred a β-hemolytic streptococcus for hemolysis and reaction to the bacitracin disk (inhibition of growth indicates *S. pyogenes*).

Observe the blood agar plate with the oxidase-positive inoculum. Gram stain well isolated colony and inoculate a set of cystine Trypticase agar (CTA) deeps containing glucose, maltose, sucrose and lactose and a nitrate semi-solid deep with a gram-negative diplococcus. Record the results of the differential disks. Incubate the cultures at 37°C.

4th day: Compare the reactions of the CTA tubes with the table on the following page. A positive reaction would be a yellow top indicating acid production. The nitrate test is developed by adding dimethylnaphthylamine and sulfanilic acid.

Neisseria

	Colony of BA (yellow pigment)	gluc	malt	sucr	lact	NO_3
N. gonorrhoeae	-	+	-	-	-	-
N. meningitidis	-	+	+	-	-	-
N. lactamicus	+	+	+	-	+	-
N. sicca	±	+	+	+	±	-
N. mucosa	±	+	+	+	-	+
N. flavescens	+	-	-	-	-	-
N. subflava	+	+	+	±	±	-
Branhamella catarrhalis	-	-	-	-	-	+

EXPERIMENT 36

THE ACID-FAST STAIN

The acid-fast stain is specific for bacteria in the order Actinomycetales. Many of these bacteria retain the stain, carbol fuchsin, when treated with 3% HCl. They are referred to as "acid-fast" bacteria. A more powerful decolorizing agent, acid-alcohol (3% HCl + 95% ethanol), is used clinically to identify *Mycobacterium tuberculosis*; the bacterium that causes human tuberculosis. These organisms retain the stain after being treated with acid-alcohol and are referred to as "acid-alcohol-fast" bacteria.

Procedure:

A. Acid-fast Bacteria

1. Prepare smears of *Staphylococcus aureus* and *Mycobacterium smegmatis* or *Mycobacterium phlei* (the mycobacteria cultures should be at least 4 to 7 day-old slant cultures). Allow smears to air dry, then heat fix them.

2. Cover the smears with carbol fuchsin and steam gently for 3 to 5 minutes; without allowing the stain to evaporate. This may be accomplished by holding the base of a Bunsen burner and tilting it, so that the flame directly warms the stain until vapors start coming off. Intermittently heat the stain for the time period, adding more carbol fuchsin if needed.

3. **Allow the slide to cool!** Rinse the stain from the slide,

4. Decolorize with a dropper full of 3% HCl.

5. Rinse with water.

6. Counter stain with methylene blue for 30 to 60 seconds.

7. Rinse and carefully blot dry. Observe under oil immersion.

Questions:

1. What other characteristics are common to acid-fast bacteria?

2. Does methylene blue stain acid-fast bacteria?

3. Is carbol fuchsin an acidic or basic stain?

EXPERIMENT 37

INTESTINAL FLORA

The intestinal tract of humans contains a great variety of microorganisms, many of which are normal residents. Organisms that survive passage through the acid of the stomach -- often protected by the food with which they are ingested -- can become established for greater or lesser time, either enjoying the bounty provided by the intestinal contents or as actual parasites, living on the membranes of the intestinal tract. Some of the latter may produce disease (e.g. cholera, bacillary dysentery, salmonellosis). The resident flora of each persons' intestinal tract is usually quite stable, and certain bacteria are generally found in residence; witnessed by their presence in feces. This exercise will be performed to demonstrate that even though the latter statement is true, the relative amounts of certain bacteria will vary from person to person.

Procedure:

1st Day. Using a sterile swab, obtain a specimen of your fecal matter. Heavily Inoculate a plate of XLD (xylose lysine deoxycholate) agar with the swab, then streak with a sterile loop. Lightly swab a plate of EMB (eosin Y - methylene blue) agar, then use a sterile loop to streak for isolation. Place the used swab into a tube of nutrient broth, then incubate the tube and the plates at 37°C for two days.

2nd Day. Examine the plates. If they have no growth, re streak the XLD and EMB plates with the swab in the nutrient broth tube. If growth is present, pick at least two colonies to TSI and nutrient agar slants. (1) From EMB; colonies of normal coliform bacteria are either black with a green metallic sheen (*E. coli*) or large pink mucoid (E. aerogenes); potential pathogens (i.e. nonfermenters of lactose) are white, grayish or translucent colonies. (2) On XLD, normal coliforms are generally inhibited, but those that do grow, produce yellow around the colonies. Pathogenic species produce red around the colonies. Some colonies may have black centers, indicating H_2S. Incubate slants at 37°C.

<u>3rd Day.</u> Make smears from the nutrient agar slants and gram stain them. If they are gram-negative cocco-bacilli, record the results of the TSI. Typical reactions are as follows: (A = acid, indicated by yellow color; NC = no change; K= alkaline, indicated by a red color)

Butt	Slant	H_2S	
AG	A or NC	-	Normal Coliforms
AG	NC/A	±	*Proteus/Providencia* species*
AG	K	±	*Salmonella* species*
A	K	+	*Salmonella typhi*
A	K	-	*Shigella* species

* To differentiate between *Proteus* and *Salmonella* inoculate urea broth. *Salmonella* will not hydrolyze urea. If it appears that you have a *Shigella*, do a motility stab tube (MIO). If motile the organism is not a *Shigella* but probably an atypical coliform.

EXPERIMENT 38

PEPTONIZATION AND FERMENTATION OF MILK

Because its contents of proteins, carbohydrates and minerals; milk provides an excellent culture medium for many kinds of bacteria. The proteins may be coagulated, either by a rennet-like enzyme, or by acid. They may be hydrolyzed, possibly with liberation of ammonia. The carbohydrates may be fermented and yield acid and sometimes gas. Litmus is added to milk to serve as an acid-base indicator or to indicate an anaerobic environment by turning colorless when reduced in oxidation-reduction reactions.

Procedure:

1. Inoculate tubes of litmus milk with E. coli, B. subtilis and S. lactis or sour cream. Include also any unknowns you are studying.

2. Incubate at 30°C for one week.

3. After two or three days of incubation, record for each tube the following in tabular form.

    a) Reaction (acid or alkalai)

    b) Presence or absence of curd

    c) Peptonization (dirty discolorization, usually beginning at the top of the medium and progressing throughout the tube)

    d) Gas fomation (bubbles on top, or deep scorings a \loong the side of the curd)

    e) Reduction (decolorization) of the litmus reagent

Questions:

1. Distinguish between acid curdding and rennet curdding.

2. What is the chemical nature of peptonization of milk?

3. What causes reduction of the indicator when certain organisms are used? Why does the color return as the culture gets older?

4. What is whey?

5. What kind of curd does *B. subtilis* produce? *S. lactis* or sour cream organisms?

Litmus Milk Reactions

Reaction	Activity of Organisms Responsible for Reaction	Appearance
rennet curd	production of rennet enzyme	soft coagulum (soft curd) with no pH change - blue
Acid	fermentation of lactose to organic acids	Pink color
Acid and Gas	fermentation of lactose to organic acids with gas	Pink, with gas bubbles accumulated at the surface
Acid Curd	fermentation of lactose to organic acids, but pH is lowered to 4.3, so casein coagulates	Pink top with firm coagulum "hard curd"
Reduction	litmus reagent acts as an H_2 acceptor	loss of blue, starts at the bottom of the tube and works its way up
Peptonization	organism produces proteolytic enzymes which digest the casein to peptones and amino acids	watery, no longer opaque like milk; either of two colors: (1) amber or straw colored if reduced, (2) bluish if oxidized
Alkaline	deamination of amino acids	blue or dirty discoloration either with or without peptonization
Inert	no growth	no change

EXPERIMENT 39

COUNTS OF MILK BACTERIA

The number of bacteria in milk is of great sanitary significance. It provides some indication of the conditions of production, handling and storage. It is also used for part of the basis for grading milk. Grade A pasteurized milk should contain not over 30,000 bacteria per ml (standard plate count) before delivery.

Procedure:

1. Samples of raw and pasteurized milk will be provided. Work with one of these.

2. Shake the sample bottle vigorously at least 25 times.

3. With a sterile 1 ml pipette, transfer 1 ml of sample to a 9 ml water blank. Mix by rotating the tube. This makes a 1:10 dilution.

4. With a second 1 ml pipette transfer 1 ml of the 1:10 dilution to (a) a sterile Petri dish, and (b) another 9 ml water blank. Mix. This gives a 1:100 dilution.

5. With a third 1 ml pipette, transfer 1 ml the 1:100 dilution to (a) another sterile Petri dish, and (b) another 9 ml water blank. Mix. This provides a 1:1,000 dilution.

6. With another 1 ml pipette, transfer 1 ml of the 1:1,000 dilution to a third sterile Petri dish.

7. Pour melted and cooled tryptone-glucose-yeast extract agar into each Petri dish. Mix the agar and milk dilutions by **carefully** rotating the dishes.

8. Incubate the plates at 30°C until next period.

9. Count the colonies, including small sub-surface colonies, on the plate that has between 25 and 250 colonies.

10. Calculate the number of bacteria per ml of sample (the standard plate count) by multiplying the number of colonies by the reciprocal of the dilution counted.

11. Record your results and also those of your neighbor.

Questions:

1. Why should milk samples be vigorously shaken?

2. What are some errors associated with the plate count?

3. How did your results compare with those of other students who used the same sample?

4. Why is a series of dilutions used when only one plate is to be counted?

5. Why is it preferable to count a plate with 25 to 250 colonies?

6. What are the important sources of bacteria in milk?

7. What effect would an incubation temperature of 35°C instead of 30°C have on the plate count? Explain. Which temperature is preferred for testing milk?

EXPERIMENT 40

COLIFORM ANALYSIS OF DRINKING WATER

Water containing intestinal bacteria is unsafe for drinking because organisms may be intestinal viruses and bacteria that cause typhoid fever, dysentery, cholera and other intestinal diseases. The total bacterial count is considered less important. This experiment consists of (1) a standard procedure used to determine the sanitary quality of water, (2) the determination of the most probable number (MPN) of total coliform bacteria, fecal coliform bacteria and *E. coli* in a water sample. The MPN method is used mainly in food analysis where food particles interfere with filtration methods.

Procedure: Presumptive Test

Day 1

1. Shake the water sample bottle 25 times.
2. With a sterile 10 ml pipette, transfer 10 ml of the sample to three large fermentation tubes containing lactose-lauryl sulfate-tryptone broth (LTB).
3. With a sterile 1 ml pipette, transfer 1 ml of the sample to three small LTB tubes and 0.1 ml to three small LTB tubes.
4. Incubate tubes at 35°C. Observe at 24 and 48 hours.
5. Observe the fermentation tubes for presence of gas. Record results.

Day 2. Confirmed Total Coliform Test

1. Select tubes that show gas or heavy growth within 24 hours or 48 hours and inoculate lactose brilliant green bile broth (BGB) fermentation tubes. Use a sterile loop for this procedure. Incubate the tubes at 35°C for 25 hours.

Confirmed Fecal Coliform test

1. Inoculate E.C. media with bacteria from LTB tubes that show gas or heavy growth within 24 or 48 hours. Incubate the tubes in a 44.5°C water bath for 24 hours.

Confirmed *E. coli* test

1. Inoculate and incubate E.C. Mug (methylumbelliferyl-glucuronide) media with bacteria from the LTB tubes used to inoculate the confirmed fecal coliform test.

Day 3. Confirmed Total Coliform test

1. Record the number of positive tubes in each set and determine the MPN/100 ml of water sample.

Confirmed *E. coli* test

1. In a darkened room, expose the E.C. Mug tubes to fluorescent light and record the number of tubes that fluoresce. [Use either an *Enterobacter* sp. for control or an E.C. no Mug tube with *E. coli* plus an uninoculated E.C. Mug tube.] This completes the *E. coli* test. Calculate the MPN.

Completed Total Coliform test

1. Streak, for isolation, LES ENDO agar inocula from each tube of BGB broth that showed gas. Incubate the plates at 35°C.

Day 4. Completed Total Coliform test

1. Pick and transfer two or more colonies considered most likely typical (red nucleated colonies with or without metallic sheen) or atypical (opaque, unnucleated, mucoid pink after 24 hour incubation) of the coliform group to LTB tubes and to a nutrient agar slant.

Day 5. Completed Total Coliform test

1. Observe gas in LTB tubes.
2. Do a gram stain from 24 hour slants.

For drinking water, the presence of any fecal coliforms per 100 ml is prohibited. For growing shellfish, no more than 10% of the water samples should exceed an MPN of 43 per 100 ml for a 5 tube decimal dilution test or 49 per 100 ml for a 3 tube decimal dilution test.

MPN DETERMINATION FROM MULTIPLE TUBE TEST

Number of tubes giving Positive Reaction Out of			MPN Index per 100 ml	95 Percent Confidence Limit	
3 of 10 ml each	3 of 1 ml each	3 of 0.1 ml each		Lower	Upper
0	0	1	3	<0.5	9
0	1	0	3	<0.5	13
1	0	0	4	<0.5	20
1	0	1	7	1	21
1	1	0	7	3	23
1	1	1	11	3	36
1	2	0	11	1	36
2	0	0	9	3	36
2	0	1	14	3	44
2	1	0	15	7	89
2	1	1	20	4	47
2	2	0	21	10	150
2	2	1	28	4	120
3	0	0	23	7	130
3	0	1	39	15	380
3	0	2	64	7	210
3	1	0	43	14	230
3	1	1	75	30	380
3	1	2	120	15	210
3	2	0	93	15	380
3	2	1	150	30	440
3	2	2	210	35	470
3	3	0	240	36	1300
3	3	1	460	71	2400
3	3	2	1100	150	4800

EXPERIMENT 41

BACTERIA IN GROUND MEAT

Food may contain a variety of microorganisms, their numbers and kinds varying with the nature and history of the food. Microbial activity often produces desirable changes in flavor or texture, but if the activity of food enzymes and microorganisms is not controlled, decomposition proceeds until the material is no longer acceptable to a discriminating consumer. The variety of microorganisms encountered precludes the use of any one culture medium or method for enumeration and isolation of all types. In this exercise only bacteria able to grow aerobically at 30°C on nutrient agar will be studied.

Procedure: Part I.

1. A 1:100 suspension of ground beef will be prepared in a food blender. From this make further dilutions (1:1,000, 1:10,000, and 1:100,000) in the manner outlined in Exp. 39, and transfer 1 ml portions of each into sterile Petri dishes.

2. Pour melted-and-cooled nutrient agar into the plates. Let the agar harden and incubate at 30°C for 2 to 5 days.

3. Count the plate having between 25 and 250 colonies and determine the number of bacteria per gram of sample. Record the results obtained by all the students at your lab bench and calculate the average. (What factors contribute to the variation between individual counts.)

4. Gram stain three representative colonies and examine.

Part II. Lipase Test

1. Inoculate a nutrient agar plate containing 0.5% Crisco.

2. Observe the plate for clearing around the colonies on the plate.

Questions:

1. Explain what happens during aging of meat.

2. Do the bacteria degrade chiefly fats, proteins or non-proteinaceous constituents of ground beef? Explain.

3. What is the difference between food infection and food poisoning (intoxication)?

4. Outline a procedure for the enumeration of bacteria, yeasts, and molds from foods.

5. Is the lipase enzyme intracellular?

EXPERIMENT 42

BACTERIOPHAGE GROWTH

When bacterial viruses (bacteriophages) attack their hosts and grow, they lyse (break open) the bacteria to release the newly produced bacteriophages. This can be observed for quantitative purposes better as plaques (holes or clearings) in a lawn of bacteria grown in a soft agar overlay of the surface of the agar in a Petri dish. Each plaque is considered to be the product of a single bacteriophage. In this experiment we will determine the burst number of bacteriophage per infected cell.

Procedure:

1. Secure 10 ml of *E. coli* in tryptone broth grown up to 1-3 x 10^8 cells per ml.

2. Infect the culture with 1 x 10^9 bacteriophages per ml (about 5 bacteriophage per bacterium). Be sure to mix the tube well by using the pipette aid to blow out the the phage delivery pipette to insure good mixing.

3. Incubate in a water bath at 37°C for 5 minutes to allow adsorption of phages to the bacteria. After 5 minutes, plate the 10^6 dilution by adding 0.1 ml of the 10^5 dilution to 5 ml of melted-soft agar overlay, add 3 drops of a turbid *E. coli* culture, mix well and pour over a nutrient agar plate.

4. Using 5 ml dilution tubes, dilute infected cells by a factor of 10^6. Place 1 ml of the 10^6 dilution in a sterile tube and incubate it at 37°C (shaking it every 10 minutes to ensure adequate aeration) for 1 hour.

5. After 1 hour incubation, dilute the 10^6 dilution 1:10 and add 0.1 ml of this dilution to 5 ml of melted-soft agar overlay, add 3 drops of a turbid *E. coli* culture, mix well and pour over a nutrient agar plate.

6. After allowing the soft agar to solidify, incubate the plates for 24 hours and count the plaques on each plate.

7. Determine the average number of bacteriophage released per cell by dividing the final number of plaques per ml by the number of plaques per ml found immediately after adsorption.

EXPERIMENT 43

OXIDASE TEST

Cytochromes are heme type compounds which transport electrons to O_2 forming H_2O during respiration. In some bacteria such as *E. coli*, and other members of the family *Enterobacteriaceae*, the cytochrome which passes the electrons directly to oxygen is of the "b" type. In others, such as members of the genera *Vibrio*, *Aeromonas* and *Neisseria*, it is of the "c" type, and the enzyme-cytochrome complex is called cytochrome oxidase. The compound N, N, N', N' tetramethyl p-phenylene diamine can be oxidized from a colorless to a colored (purple) form only by bacteria which have a cytochrome c oxidase. The diagnostic or identification test based on this difference is called the "oxidase test".

Procedure:

The oxidase reagent is somewhat toxic and should be kept away from the skin. If it comes in contact with your skin, wash it off immediately. The reagent is unstable (it will auto oxidize to the colored form) and must be kept cold, and protected from light.

1. Streak a nutrient agar plate that has no glucose for isolation with a mixture of *Aeromonas hydrophila* and *E. coli*.

2. Incubate the plate at 30 °C until next period.

3. Add a drop of oxidase reagent to the surface of a colony. Continue to do this until you have found one colony which is oxidase positive (it turns dark purple) and one which is oxidase negative (clear to light purple).

4. The reagent is also toxic to bacteria. If you wanted to subculture from a colony identified as oxidase positive or negative, you would have to do so immediately.

5. An alternative way to do an oxidase test is to remove some of the growth from the colony with a toothpick to a filter paper saturated with oxidase reagent.

Handwashing Effectiveness Laboratory

Introduction:

This laboratory assignment will demonstrate the effects of three methods of hand-washing upon the numbers of microbial flora of the skin as compared to unwashed skin.

Materials:

4 plates of Trypticase soy agar or nutrient agar (one plate per person)
4 sterile swabs (per bench)
sterile water (3-5ml per tube, one tube per person)
antiseptic cleaning solution (Betadine, pHisoHex, Microquat, etc.)
sterile hand brush (2 per bench)

Procedure:

1. Divided students into groups of four & assign each student a either #1, #2, #3, or #4.

2. Have group member #1 label the four agar plates as follows:
   a. Names, date, #1, Unwashed
   b. Names, date, #2, Washed
   c. Names, date, #3, Washes and Scrubbed
   d. Names, date, #4, Washed, Scrubbed Antiseptic

3. Group member #1 is the timer.
   Group member #2 is the handwashing subject.
   Group member #3 is the swabber.
   Group member #4 is the streaker.

4. Group member #3 should dampen the sterile swab in the tube of sterile water and express the escess water from the swab by rolling it against the inner wall of the tube. The swabber (#3) should swab *a portion of the palm and small finger* of the handwashing subject's hand (#2).

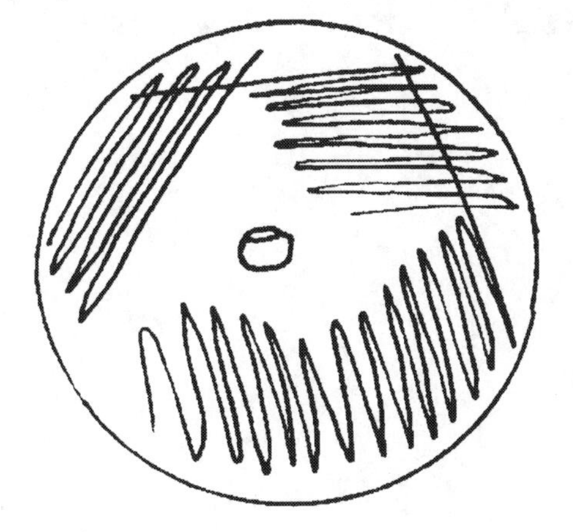

5. Group member #4 is to streak the swab across plate #1, the first sector should be streaked with the swab and the other sectors should be streaked for isolation with a sterilized and cooled loop but leaving a small area clear in the center of the plate. Group member #2 (the hand washer) should press his/her *pinkie fingertip and nail* into the agar.

6. Group member #2 washes his/her hands for 20 seconds with hot soapy water. (Group member #1 should time the 20 sec.) Using a new swab, group member #3 should swab *the ring finger and part of the palm* of group member #2. Group member #4 streaks the second plate for isolation, leaving a clear area in the center of the plate for group member #2 to press his/her *ring fingertip and nail* into the agar.

7. Next group member #2 washes his/her hands with hot soapy water and a hand brush for 20 seconds (with group member #1 timing). Then group member #2 should spend 20 seconds scrubbing the wet soapy brush on the fingernails of the hand. When finished, group member #2 should rinse his/her hands and brush thoroughly with hot water. The procedure should continue as before with the swabbing and streaking only this time *the middle finger and palm should be swabbed* and the third plate should be streaked, followed by pressing *the middle finger tip and nail* into the center of the agar plate.

8. Group member #2 washes his/her hands for 20 seconds with hot water, the brush and an antiseptic cleansing agent. Spend the next 20 seconds using the brush and cleansing agent on the fingernails. Rinse the hand. Repeat the same procedure swabbing *the index finger* this time and using plate #4 (repeat as above, with the proper fingertip).

9. A pair of students will be selected by the instructor to label 3 plates Control #1, Control #2, and Control #3. Control #1 should be swabbed for isolation using the soap. Control #2 should swabbed for isolation using a swab which has been moistened in sterile water and rubbed over a sterile hair handbrush. Control #3 should be swabbed with the antiseptic cleansing agent.

10. Incubate all plates at 30°C for one or two days.

Cleanliness and Sanitation

Materials:

3-5 ml tubes of sterile 0.85% saline
3 cotton tip swabs
3 nutrient agar plates (per pair)
1 inoculum sprayer (sneezer loaded with a dilution of *Micrococcus luteus*)
Rocal
template

A. *Aerosol Generation and Detection*

This exercise demonstrates the spread of microorganisms by aerosol generation

1. Label your plate with your name, date and the exercise performed
2. Each member of your group should place a plate of nutrient agar on the bench and remove the lid. (They should be placed in a line that will allow the last plate to be farthest away from the "sneeze")
3. Use the sprayer and "sneeze" once towards the plates
4. Replace the lids on the plates and incubate them at 30°C

B. *How Clean are Objects?*

This exercise demonstrates the ubiquitous presence of bacteria and fungi on objects

1. Each person should select an item such as a door knob, to determine whether there are microorganisms on its surface.
2. Label a nutrient agar plate with your name, date and the object sampled.
3. Select one of your swabs for this part of the exercise, remove the swab, dampen the swab in sterile saline, then clean the entire surface of the object.
4. swab the surface of your nutrient agar plate, then return the swab to its tube.
5. Incubate the plate at 30°C.

C. ***Effect of Work Area Decontamination*** (per pair)

The effects of cleaning an area with a disinfectant

1. Label two nutrient agar plates with your name, date and the experiment (one before and one after).

2. Place the template on your work area. Remove a swab from the sterile saline and swab the area inside the template.

3. Swab the surface of the before plate.

4. Pour some Rocal solution onto the work area and clean the area with a paper towel. Allow the area to dry.

5. Place the template on the same spot as before and swab the area as before with a sterile swab. Swab the after plate and incubate at 30°C.

Classroom Epidemic Laboratory

Introduction:

In this exercise, a classroom epidemic will be simulated! It will safely demonstrate how easily organisms (both pathogenic and nonpathogenic) may pass from person to person. A lollipop contaminated with baker's yeast will begin the epidemic and it will be spread throughout the class by handshakes.

Materials:

Trypticase soy agar or Sabouraud's dextrose agar plates (2 plates per bench)
sterile swabs - one per student
very active young culture of *Saccharomyces cerevisiae*
1 lollipop per team of 8 students

Procedure:

1. Each lab bench will serve as a team. The team should line up in a row.

2. Have one agar plate for each four students: sector plates into three or four parts and label with initials, team and member number and date (seven of eight will participate).

3. All team members are to wash and dry their hands thoroughly. They should not touch any contaminated surfaces until the lab is completed.

4. One instructor will dip the lollipop into the yeast suspension and "shake hands" (lollipop to hand) with student #1 of team one.

5. Student #1 of team one will now shake hands with student #2 of team one (sticky lollipop hand to hand). This will continue down the line until all members of team one have shaken hands with the student in front of them. (#1 shakes hands with #2, #2 shakes hands with #3, etc.)

6. The eighth member of the team should follow with sterile swabs and swab the right hands of each student immediately after they have shaken hands with the student in front of them. Once a students hand has been swabbed, he/she should spread it uniformly on his/her section of the plate.

7. After swabbing their section of the plate, students should wash their hands thoroughly.

8. Incubate the plates at 30°C.

Appendix A - Special Comments for Experiments

Exp. 2: Tube A should contain a coccus (*Staphylococcus epidermidis or Micrococcus luteus*) and tube B should contain a rod (*Bacillus subtilis, B. cereus or B. megaterium*). The cultures should be broth cultures, since beginning students find it very hard not to make their smears from agar surface cultures too heavy with bacteria.

Exp. 3: A mixture of *E. coli* and *Micrococcus luteus* made from individual cultures, blended just before use in the laboratory (you might substitute *Serratia marcescens* and *S. epidermidis*).

Exp. 4: The decolorization step with 95% ethanol is critical. If the smear is not too thick, the problem is usually over decolorization which may be corrected by shortening steps 4 and 5 to 5-6 seconds.

Exp. 5: Remember to use a young culture 16-24 hour broth for this experiment.

Exp. 12: Wear gloves when disposing of culture tubes containing nitrate reagents, since they are potential carcinogens.

Exp. 14: *Citrobacter* species is another H_2S positive bacterium that can be used.

Exp. 15: *Enterococcus faecalis* is a negative control for catalase.

Exp. 16: If the V.P. test for *E. aerogenes* doesn't turn pink, add a small amount of creatinine-HCl (this is a source of guanindino- group, like arginine, that allows color development).

Exp. 17: This may also be performed with *E. coli* in nutrient broth using side-arm flasks to facilitate taking OD. readings. The variables could be: (a) addition of glucose; (b) aeration (shaking); (c) incubation temperature (room temperature vs. 37°C); or (d) inoculum (log. phase vs. stationary phase).

Exp. 18: A special strain (ATCC 9986) of *Serratia marcescens* is required for this experiment.

Appendix B - Stains and Reagents

Acid for Acid-Fast Stain
3 ml of conc. HCl in 100 ml of water

Alpha-Naphthol Reagent (Voges Proskauer)
Solution A: 5 g alpha-naphthol in 100 ml ethanol
Solution B: 40 g KOH in 100 ml H_2O

Carbol-Fuchsin Stain (Acid-Fast Stain)
Solution A: 0.3 g Basic Fuchsin in 10 ml 95% ethanol
Solution B: 5 g phenol in 95 ml H_2O
Mix A & B and filter

Copper Sulfate (Capsule Stain)
20 g $CuSO_4$ in 80 ml H_2O, Heat

Crystal Violet Stain (Simple & Gram Stain)
Solution A: 2 g Crystal Violet in 20 ml of 95% ethanol
Solution B: 0.8 g ammonium oxalate in 80 ml H_2O
Mix A & B and Filter

Gelatin Developer
20 ml conc. HCl in 80 ml H_2O

Lugol's Iodine (Gram Stain)
Dissolve 2 g KI in 300 ml H_2O, add 1 g I_2

Kovac's Reagent
n-Amyl alcohol 75 ml
HCl (conc) 25 ml
p-Dimethylaminobenzaldehyde 5 g

Malachite Green (Spore Stain)

5% Solution of Malachite Green Oxalate
allow to stand 30 min. then filter

Loeffler Methylene Blue
Solution A: Dissolve 0.3 g of methylene blue in 30 ml 95% ethanol
Solution B: 0.01 g KOH in 100 ml H_2O
Mix Solutions A & B

Nigrosin
Boil 10 g water soluble nigrosin in 100 ml of H_2O for 30 min., add 0.5 ml formaldehyde (40%) as a preservative. Filter twice through double filter paper & store under aseptic conditions.

Dimethyl-1-naphthylamine (Nitrite Test)
Solution A: 0.8% Sulfanilic acid in 5 N Acetic Acid
Solution B: 0.6% NN'-Dimethyl-α-naphylamine in 5 N Acetic Acid

Note: Dimethylnaphthylamine is a carcinogen. Avoid contact with skin. Wear face mask when working with powder.

Oxidase Reagent
Solution A: 0.6 g NN'-Dimethyl-p-phenylene-diamine oxalate in 60 ml H_2O. Heat slightly, then cool immediately.
Solution B: 40 ml of 1% α-naphthol
Make Reagent Fresh, keep on ice.

Safranin (Gram Stain)
10 ml of 2.5% Safranin O in 95% ethanol to 100 ml of H_2O, Filter

Appendix C - Culture Media

Blood Agar

Defibrinated Blood	50 ml
Beef Heart Infusion	500 g
Tryptose	10 g
Sodium Chloride	5 g
Agar	15 g

Bacto Lauryl Tryptose Broth

Bacto tryptose	20 g
Lactose	5 g
Potassium Phosphate dibasic	2.75 g
Potassium Phosphate monobasic	2.75 g
Sodium Chloride	5 g
Sodium Lauryl Sulfate	0.1 g

Casein Agar

Nutrient Agar plus
10% skim milk powder

Citrate Agar (Simmons)

Magnesium Sulfate	0.2 g
Monoammonium Phosphate	1 g
Dipotassium Phosphate	1 g
Sodiium Citrate	2 g
Sodium Chloride	5 g
Agar	15 g
Brom Thymol Blue	0.08 g

Plate Count Agar

Yeast Extract	2.5 g
Tryptone	5 g
Glucose	1 g
Agar	15 g

E.M.B. Agar

Peptone	10 g
Lactose	10 g
Dipotassium Phosphate	2 g
Agar	15 g
Eosin Y	0.4 g
Methylene Blue	0.065 g

MRVP Broth

Buffered Peptone	7 g
Glucose	5 g
Dipotassium Phosphate	5 g

Litmus Milk

Fresh Skim Milk	1000 ml
Litmus, powdered	5 g

Mannitol Salts Agar

Beef Extract	1 g
Proteose Peptone No. 3	10 g
Sodium Chloride	75 g
d-Mannitol	10 g
Agar	15 g
Phenol Red	0.025 g

Nitrate Broth

Beef Etract	3 g
Peptone	5 g
Potassium Nitrate	1 g

Peptone Iron Agar

Peptone	15 g
Proteose Peptone	5 g
Ferric Ammonium Citrate	0.5 g
Dipotassium Phosphate	1 g
Sodium Thiosulfate	0.08 g
Agar	15 g

Tryptone Broth

Tryptone	10 g
Beef Extract	3 g
Sodium Chloride	5 g

Urea Broth

Yeast Extract	0.1 g
Monopotassium Phosphate	9.1 g
Dipotassium Phosphate	20 g
Urea	20 g

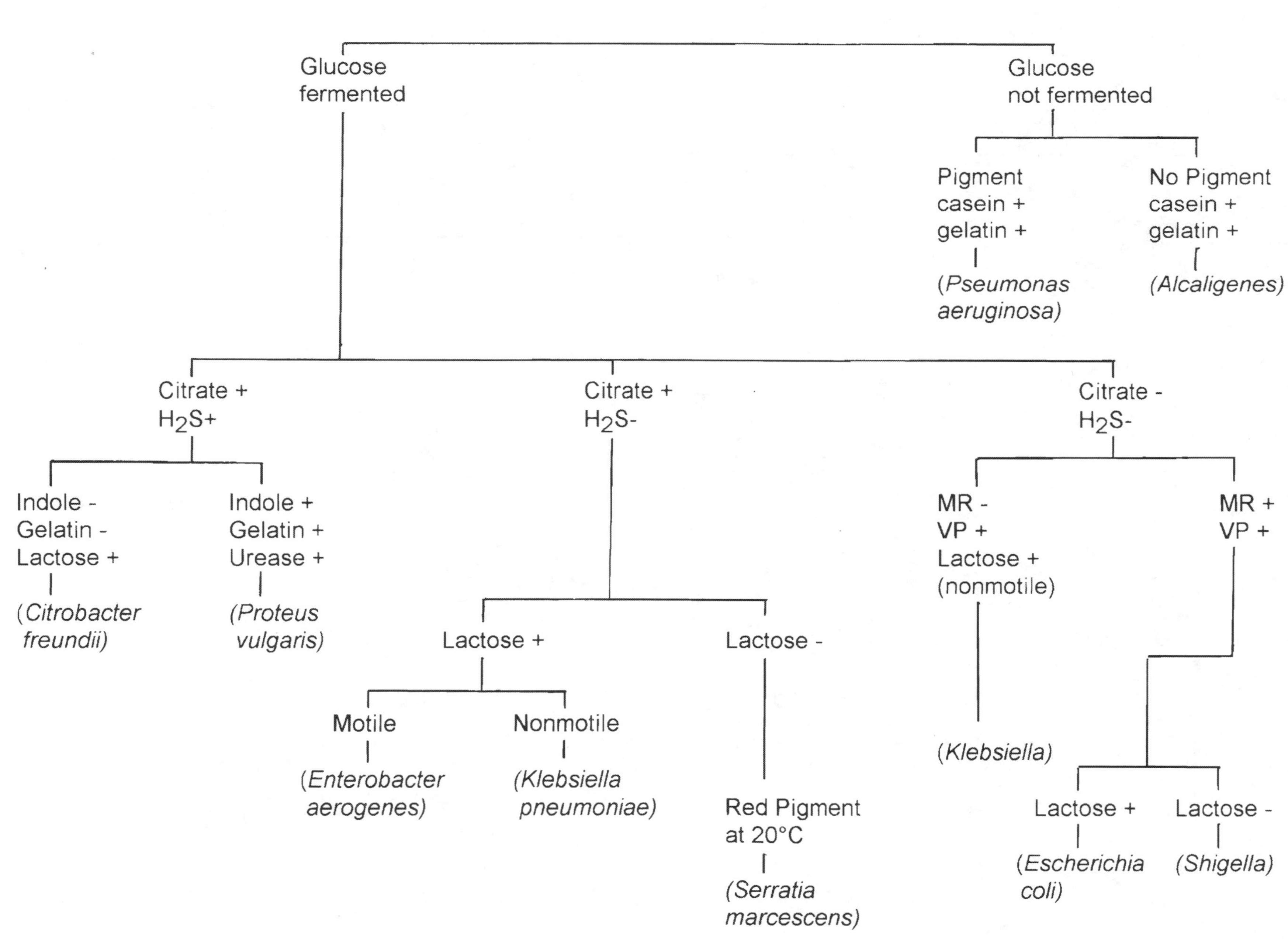

Key to Selected Gram Negative Rods
Glucose fermented
Glucose not fermented
Pigment casein + gelatin +
(Pseumonas aeruginosa)
No Pigment casein + gelatin +
(Alcaligenes)
Citrate + H2S+
Citrate + H2S-
Citrate - H2S-
Indole - Gelatin - Lactose +
(Citrobacter freundii)
Indole + Gelatin + Urease +
(Proteus vulgaris)
Lactose +
Lactose -
Motile
(Enterobacter aerogenes)
Nonmotile
(Klebsiella pneumoniae)
Red Pigment at 20°C
(Serratia marcescens)
MR - VP + Lactose + (nonmotile)
(Klebsiella)
MR + VP +
Lactose +
(Escherichia coli)
Lactose -
(Shigella)

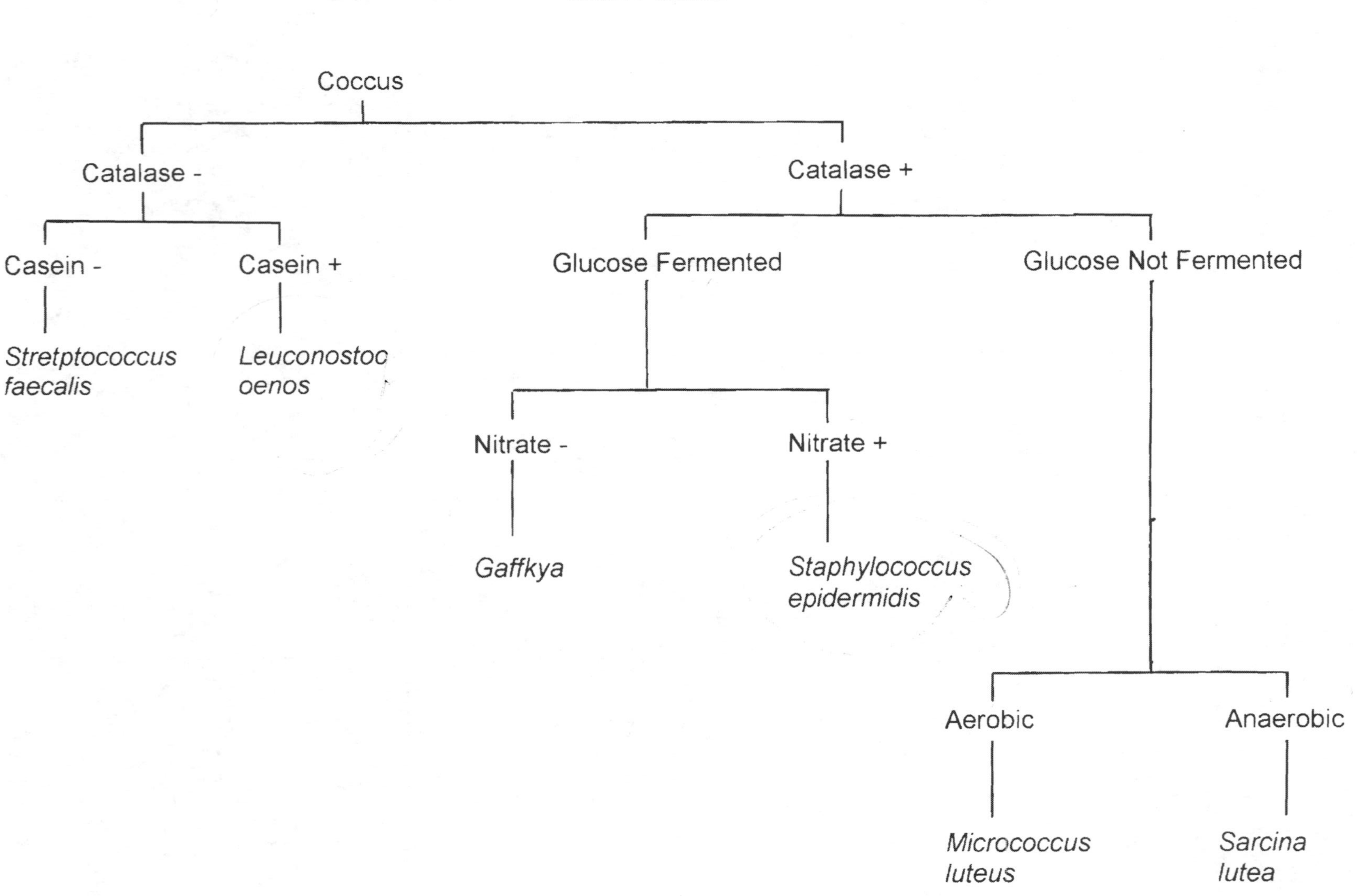
Gram Positive
Coccus
Catalase -
Catalase +
Casein -
Casein +
Stretptococcus faecalis
Leuconostoc oenos
Glucose Fermented
Glucose Not Fermented
Nitrate -
Nitrate +
Gaffkya
Staphylococcus epidermidis
Aerobic
Anaerobic
Micrococcus luteus
Sarcina lutea

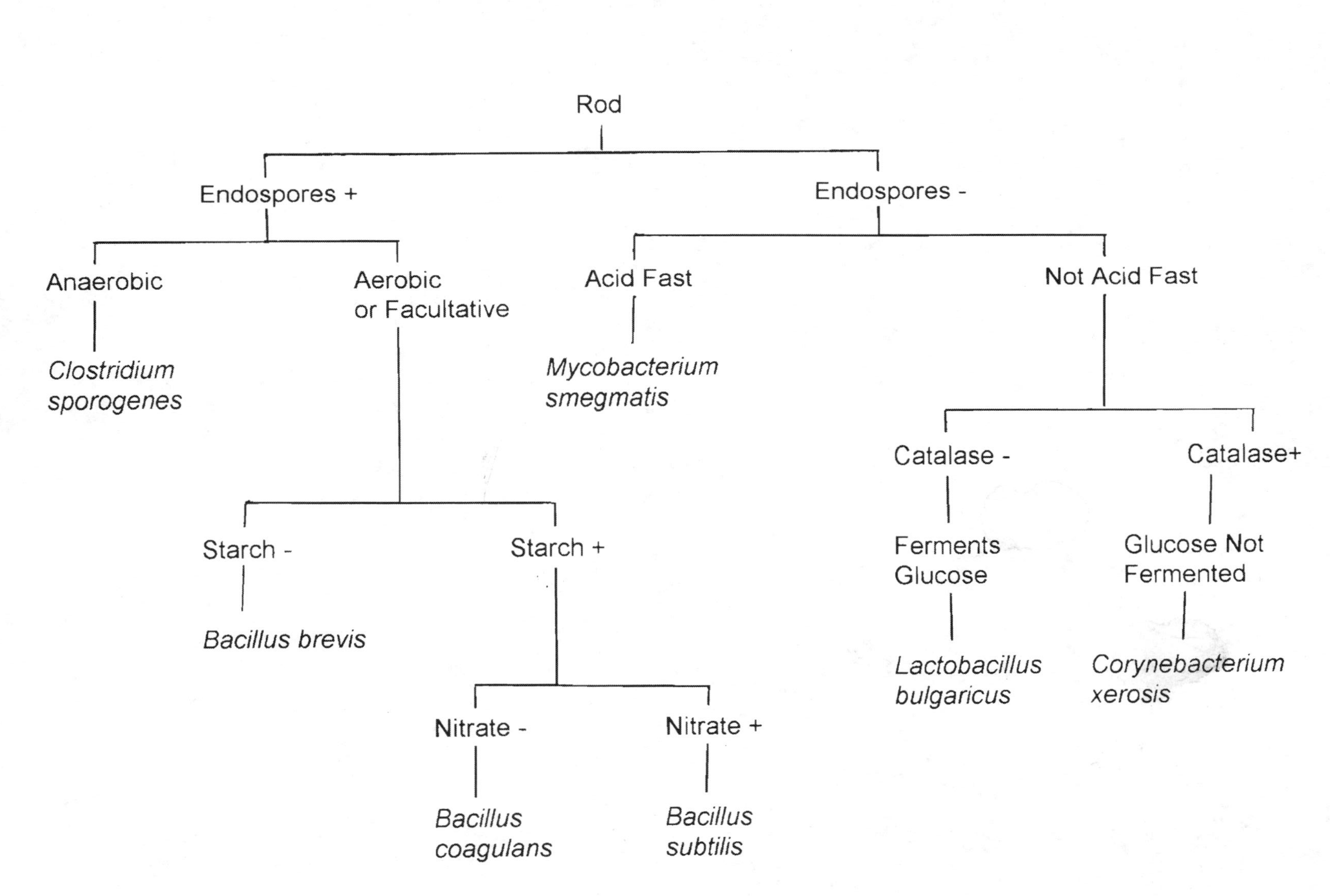

Gram Positive Bacteria
Rod
Endospores +
Endospores -
Anaerobic
Aerobic or Facultative
Acid Fast
Not Acid Fast
Clostridium sporogenes
Mycobacterium smegmatis
Catalase -
Catalase+
Starch -
Starch +
Ferments Glucose
Glucose Not Fermented
Bacillus brevis
Lactobacillus bulgaricus
Corynebacterium xerosis
Nitrate -
Nitrate +
Bacillus coagulans
Bacillus subtilis